Informing Water Policies in South Asia

This cluster of books presents innovative and nuanced knowledge on water resources, based on detailed case studies from South Asia–India, Bangladesh, Bhutan, Nepal, Pakistan, and Sri Lanka. In providing comprehensive analyses of the existing economic, demographic and ideological contexts in which water policies are framed and implemented, the volumes argue for alternative, informed and integrated approaches towards efficient management and equitable distribution of water. These also explore the globalization of water governance in the region, particularly in relation to new paradigms of neoliberalism, civil society participation, integrated water resource management (IWRM), public–private partnerships, privatization, and gender mainstreaming.

Water Resources Policies in South Asia
Editors: Anjal Prakash, Sreoshi Singh, Chanda Gurung Goodrich and S. Janakarajan
ISBN 978-0-415-81198-9

Globalization of Water Governance in South Asia
Editors: Vishal Narain, Chanda Gurung Goodrich, Jayati Chourey and Anjal Prakash
ISBN 978-0-415-71066-4

Informing Water Policies in South Asia
Editors: Anjal Prakash, Chanda Gurung Goodrich and Sreoshi Singh
ISBN 978-0-415-71059-6

Water Governance and Civil Society Responses in South Asia
Editors: N. C. Narayanan, S. Parasuraman and Rajindra Ariyabandu
ISBN 978-0-415-71061-9

Informing Water Policies in South Asia

Editors

Anjal Prakash
Chanda Gurung Goodrich
Sreoshi Singh

First published 2014 in India
by Routledge
912 Tolstoy House, 15–17 Tolstoy Marg, Connaught Place, New Delhi 110 001

Simultaneously published in the UK
by Routledge
2 Park Square, Milton Park, Abingdon, Oxon OX14 4RN

Routledge is an imprint of the Taylor & Francis Group, an informa business

Typeset by
Glyph Graphics Private Limited
23, Khosla Complex
Vasundhara Enclave
Delhi 110 096

British Library Cataloguing-in-Publication Data
A catalogue record of this book is available from the British Library

ISBN 978-0-415-71059-6

Contents

List of Tables

List of Figures

List of Maps and Boxes

Maps

Boxes

List of Abbreviations

ABECA	All Bengal Electricity Consumers Association
ADB	Asian Development Bank
AWB	Area Water Board
BDO	Block Development Officer
BoM	Board of Management
CAT	Convention Against Torture
CBO	Community-Based Organization
CEDAW	Convention on Elimination of All Forms of Discrimination Against Women
CEI	Composite Empowerment Index
CERD	Convention on Elimination of All Forms of Racial Discrimination
CGWB	Central Ground Water Board
CIDA	Canadian International Development Agency
CMSU	Community Management Support Unit
CPR	Common Property Resource
CRC	Convention on the Rights of the Child
CSO	Civil Society Organization
CST	Cluster Storage Tank
DPC	District Planning Committee
DR & WCS	Drainage & Water Conservation
DSL	dead storage level
DSC	Development Support Centre
EI	Empowerment Index
ERR	Earthquake Reconstruction and Rehabilitation
FCD	Flood Control and Drainage
FGD	focus group discussion
FO	farmers' organizations
FRL	full reservoir level
GEM	Gender Empowerment Measure
GGA	Groundwater Governance in Asia
GIDR	Gujarat Institute of Development Research
GLOFs	glacier lake outburst floods
GMS	Greater Mekong Sub-basin

GoI	Government of India
GoM	Government of Maharashtra
GoN	Government of Nepal
GoWB	Government of West Bengal
GR	government resolution
GRWSSP	Ghogha Rural Water Supply and Sanitation Project
GWA	Gender and Water Alliance
GWF	Global Water Forum
GWP	Global Water Partnership
GWSSB	Gujarat Water Supply and Sewerage Board
HUDA	Haryana Urban Development Authority
HYV	high-yielding variety
ICCPR	International Covenant on Civil and Political Rights
ICESCR	International Covenant on Economic, Social and Cultural Rights
IDRC	International Development Research Centre
IGA	income-generating activities
IIMI	International Irrigation Management Institute
ILO	International Labour Organization
IMTP	Irrigation Management Transfer Project
I-NRM	Integrated Management of Natural Resources
IPD	Irrigation and Power Department
ISA	Implementation Support Agency
IWE	Irrigation and Water Engineering
IWMI	International Water Management Institute
IWRM	Integrated Water Resources Management
IWT	Indus Water Treaty
KMVS	Kutch Mahila Vikas Sangathan
LCs	local committees
LGED	Local Government Engineering Department
MDG	Millennium Development Goal
MGD	Million Gallons per day
MMISF	Maharashtra Management of Irrigation Systems by Farmers
MoU	Memorandum of Understanding
MPA	Megh Pyne Abhiyan

MPCs	Municipal Planning Committees
MSPs	Multi Stakeholder Platforms
NAPA	National Adaptation Plan of Action
NGO	Non-Governmental Organization
NPA	National Plan of Action
NPDEW	National Policy for Development and Empowerment of Women
NSA	Network for Social Accountability
NWSDB	National Water Supply and Drainage Board
O&M	Operations and Maintenance
OECD	Organisation for Economic Co-operation and Development
OLS	Ordinary Least Square
OWPO	Orissa Water Planning Organization
PGCIL	Power Grid Corporation of India Limited
PHED	Public Health and Engineering Department
PIM	Participatory Irrigation Management
PRIs	Panchayati Raj Institutions
PRSP	poverty reduction strategy paper
PUI	Peri-urban Interface
RA	Regulatory Authority
RBA	River Bed Aquifer
RWAs	Resident Welfare Associations
SAP	Structural Adjustment Program
SEB	state electricity board
SEWA	Self Employed Women's Association
SEZ	Special Economic Zone
SHGs	Self Help Groups
SIDA	Sindh Irrigation and Drainage Authority
SISP	Second Irrigation Sector Project
SOPPECOM	Society for Promoting Participative Ecosystem Management
SSWRDSP	Small Scale Water Resources Development Subprojects
SURGE	Society for Urban Regeneration
SWMO	Sindh Water Management Ordinance
TISS	Tata Institute of Social Sciences
TNAU	Tamil Nadu Agricultural University
TOD	Time of the Day

TRARC	Tropical Rapid Appraisal of Riparian Condition
UA	Urban Agglomeration
UNDP	United Nations Development Programme
UNFCCC	United Nations Framework Convention on Climate Change
UN-INSTRAW	United Nations International Research and Training Institute for the Advancement of Women
UNFPA	United Nations Fund for Population Activities
VWSC	Village Water and Sanitation Committees
WASMO	Water and Sanitation Management Organisation
WATSAN	water supply and sanitation
WB	World Bank
WBSEB	West Bengal State Electricity Board
WBSEDCL	West Bengal State Electricity Distribution Corporation Limited
WBSERC	West Bengal State Electricity Regulatory Commission
WCS	Water Conservation
WDP	Watershed Development Program
WED	Women Environment and Development
WEDO	Women's Environment & Development Organization
WEI	Women's Empowerment Index
WEM	Water Extraction Mechanism
WFG	Women Farmers' Group
WLD	Water Lifting Device
WMCA	Water Management Cooperative Association
WUA	Water Users' Association

Foreword

South Asian region has the largest population density in the world and houses around 21 percent of the world's population. A majority of this population lives below poverty line and draws sustenance from water and environmental systems. Water resources in the South Asian region has been characterized by growing stress due to competing demands that have resulted in mounting conflicts over access to and distribution of water resources and have thrown up several challenges for better water management in the region. The crisis of water management in South Asia is not divorced from the issues of governance and policy-making. Water policies in the region have been criticized for not representing the challenges people face on the ground. This book brings together local case studies from Bangladesh, India, Pakistan and Nepal. These case studies, ranging from issues of surface and groundwater irrigation to rural, urban and peri-urban water management to water supply and sanitation have raised some important concerns about participatory water governance. These issues influence water policies in more than one way. First, they tell us that the problem of water management needs to be urgently solved not only for the sustainability of environmental services, but also for the political and economic stability of individual governments in South Asia. Second, they pose some still unanswered questions on the status of institutional mechanisms for protection, conservation and sustainable use of basic environmental resources such as water. Third, all the case studies call for integration for the purpose of better water management in the region.

This book is part of the project called "Crossing Boundaries: Regional Capacity Building on Integrated Water Resources Management (IWRM) and Gender & Water in South Asia." The project aims to contribute to a paradigm shift in water resources management in South Asia, by means of a partnership-based program for the capacity-building of water professionals on IWRM and Gender & Water. The project is being implemented by six partner institutions in Bangladesh, India, Nepal and Sri Lanka, and is coordinated by the South Asia Consortium for Interdisciplinary Water Resources Studies (SaciWATERs), in partnership with the Irrigation and Water Engineering (IWE) Group, Department of Environmental Sciences, Wageningen University and Research Center, the Netherlands. The project is supported by the Ministry of Development Cooperation, Directorate-General of International Cooperation (DGIS), Government of the Netherlands. As a part of the project, SaciWATERs organized an international conference on "Water Resources Policy in South Asia" in Colombo, Sri Lanka, during December 17–20, 2008. The conference brought together many regional and international water professionals, academicians, policy-makers, activists, politicians and others involved and interested in South Asian water resources issues. It ventured out to enhance the understanding of water resources governance, and water management and use in the region by critically assessing the ongoing policy reform processes.

The chapters appearing in this book were presented as papers, discussed and deliberated at the international conference. Later, they were peer-reviewed and brought to the present shape, covering nuanced case studies on topical issues such as gender and water, groundwater, water conflict and cooperation and emerging challenges for water management and policies in South Asia. This book is part of a series of readers produced under the Crossing Boundaries Project. Largely written by South Asian scholars, they dwell on water resources issues and different aspects of IWRM in South Asia. At present, there are 10 books underway as part of the initiative.

I hope that the present book will be of interest to all those concerned with water resource management and policy-formulation process in South Asia.

E. R. N. Gunawardena
Professor, Department of Agricultural Engineering
University of Peradeniya
Sri Lanka

Acknowledgements

This book is a product of long association and engagement with scholars working on South Asian water issues. Most of the chapters that form this book were first presented at the International Conference on Water Resources Policies in South Asia held in December 2008 at Colombo, Sri Lanka. Financial grant from the Government of The Netherlands is gratefully acknowledged for holding the conference and supporting the travel and stay of the paper contributors. The contributors have been extremely patient, responding to all queries by the editors, at every stage of the publication process. However, a special thank you to the research staff and the administration at SaciWATERs, who have assisted during the organization of this conference and subsequently in shaping this volume which brings out case studies across south Asia. The editors acknowledge the comments and suggestions made by the participants of the International Conference on Water Resources Policy in South Asia that has improved the chapters to a great extent.

The editors recognise the support from the external reviewers who reviewed some of these papers toimprove its substance. We greatly acknowledge the contributions from Dr. Ajit Menon of Madras Institute of Development Studies, Chennai, India, Prof. Suresh Raj Chalise, Government of Nepal, Kathmandu, Nepal, Professor Nina Laurie, University of Newcastle, United Kingdom, Professor H S Shylendra, Institute of Rural Management, Anand, India, Professor Vishwa Ballabh, Xaviour Labour Relations Institute, Jamshedpur, India, Dr. Renu Khosla, Centre for Urban and Regional Excellence, New Delhi, India, Dr. Debolina Kundu, National Institute of Urban Affairs, New Delhi, India, Dr. Priyanie Amerasinghe, International Water Management Institute (IWMI), Hyderabad, India, Mr. Ramaswamy R. Iyer, Centre for Policy Research, New Delhi and Dr. Sunder Subramanian, formerly with Development Consulting ICRA Management Consulting Services Limited, Gurgaon, India who anonymously reviewed the papers and gave wide-ranging observations to further improve the content of this book. Support from Dr. Dibya Ratna Kansakar, former Executive Director of SaciWATERs is acknowledged for reviewing some papers of this book in its earlier form.

Professor S. Janakarajan, Madras Institute of Development Studies, India, Professor Peter Mollinga, School of Oriental and African Studies, UK and Professor Nimal Gunawardena, University of Peradeniya, Sri Lanka is acknowledged for their overall support in this project.

At the end, the editors are indebted to all those who directly or indirectly facilitated the production of this book.

PART I

Understanding of Need of Integrated Water Management in South Asia

Informing Water Policies for Integrated Water Resources Management in South Asia

Anjal Prakash, Chanda Gurung Goodrich and *Sreoshi Singh*

Gross mismanagement and exploitation of water resources have been reflected in increasing conflict between different users and uses of water. This process is coupled with extensive unplanned growth of urban centers, unsustainable exploitation of natural resources, uncontrolled industrialisation, increasing water demand for food production and expanding populations – all of which are putting pressure on limited freshwater resources. The need for finding new and innovative approaches to address the problem has led to the development of the concept of Integrated Water Resources Management (IWRM), which shows the way to move away from the earlier sub-sector approach to a more holistic or integrated approach to water management. This approach is taken up within the context of increasing scarcity (water stress), inter-sectoral conflicts, pollution and lack of technical understanding on issues of water catchments. The attention to sustainable water resource management bears testimony to an increasing sensitivity to water resource issues at a theoretical scale, and IWRM provides a conceptual framework and tool to mitigate past abuse and ensure the sustainability of water resources in future (Cap-Net 2005).

While the concept of IWRM is well defined and also debated, there are fewer case studies that inform the nitty-gritty of integration that happens at the grassroots level. This book fills the gap, as it carries case studies from the South Asian region and advocates for a much more integrated management of water resources. In particular, the issues of IWRM are viewed in the context of gender and water, groundwater challenges, water conflicts and newer contexts that inform policy-formulation processes around water. The book is divided into five sections. Section One introduces the subject to the readers, seeking to give them a conceptual understanding of the need for integrated water management in South Asia. Section Two looks into the issue of the gendered nature of participation in water management in Bangladesh, Pakistan and Nepal, through case studies in these countries. Section Three deliberates on groundwater management issues. Section Four documents water conflicts involving one or more South Asian countries, and calls for better inter-governmental cooperation in the interest of integrated water management. Section Five looks at the water issues where the contexts are fast changing. The issues raised are those of peri-urban areas, recognition of local water knowledge and co-ordination between water users and uses. Overall, the case studies in this volume bring forth critical and grounded perspectives on the issues of water management that will hopefully inform future water policies for better water management in South Asia.

From Sectoral to Integrated Management of Water

In the wake of water scarcity, climate vulnerabilities and mounting environmental degradation, water issues are becoming more complex, posing newer challenges for better governance and management. In the mainstream perspectives, water has been traditionally looked as a resource without much heed given to the inherent governance and institutional factors that determine access to and control over water. The larger goal of development is achieved through governance and policy/institutional frameworks. These frameworks are translated into action through policies seeking to achieve the goals of ecological sustainability, economic efficiency and social equity (Agarwal et al. 2000). These goals are also called the three pillars of IWRM. However, under the sectoral approach to water management, the sectors, viz., irrigation, domestic consumption and industrial use, do not coordinate with each other and work independently. They produce positive or negative externality which is absorbed by the environment or the people lowest in the power and social hierarchy and has negative implications for all the three pillars of IWRM.

In the coordinated and integrated approach, following IWRM principles, all sectors are aligned, whereby the assessment of resource availability, use and analysis, etc., are done according to the priorities defined for each of the sectors. Each sector has its own set of priorities that are collectively defined and have a combined impact on the resources. Integration is needed at the levels of policy-formulation, planning and implementation right from the grassroots through the middle and to the top management. IWRM looks into what impact each sector has on water resources and the ways to minimize that impact so that we can work towards better and safer environment (Solanes and Gonzalez-Villarreal 1999). As countries move towards a water-scarcity situation, integration becomes much more critical because when a resource is abundant there is not much problem, and it is only scarcity that brings out the worst of management, especially in case of water supply. Cities in South Asia are examples of these problems.

IWRM, therefore, requires a system of policies, laws, regulations, platforms and mechanisms to support its activities and players. It requires a culture that facilitates and encourages communication and participation of all stakeholders, particularly marginalized groups which tend to be the most adversely affected and yet have the least say in decisions on water resources management (Asaf 2010). However, critiques contend that the principles of IWRM may not be applicable to a region as diverse as South Asia. Biswas (2004) argues that the application of IWRM to more efficiently managed macro-and meso-scale water policies, programs and projects has been dismal. The systems of governance, legal frameworks, decision-making processes, and types and effectiveness of institutions often differ from one country to another in very significant ways, which a single paradigm of IWRM has not been able to encompass. Implementing IWRM also requires blending of knowledge from different disciplines that can bring valuable information about the possible consequences of decisions and actions for implementing and intended solution. Interdisciplinary approach is, therefore, essential for laying the scientific foundations for solving complex problems that need integration across the disciplines of biological, physical and social sciences in a very broad sense. Bandyopadhyay (2009) discusses the lack of institutional incentives as a root cause for the inability of policy-makers, water professionals, water rights activists and academics to cross disciplinary boundaries and to question the status quo of present water management thinking and practice. He expects that the institutional mechanisms for facilitating production and dissemination of newer knowledge for water would feed into the real world and be used for fulfilling the larger cause of poverty alleviation.

This introductory chapter is followed by the chapter by Amita Shah and Anjal Prakash that outlines the present debates around the implementation of IWRM in South Asia. The focus on IWRM discourse in South Asia has shifted from a supply-driven approach to a demand-responsive one in many countries of South Asia. The mainstream definition of IWRM has been critiqued for being ambiguous and imprecise, and demands have been made for turning it into a process and people-oriented approach to sustainable development of water resources. The debate so far has been polarized between rejecting the existing approach as part of the neoliberal agenda of giving free rein to privatization of water resources and questioning its relevance to developing countries on the one hand, and accepting the notion of IWRM as politically benign and multilayered on the other. This chapter sup-ports a third approach while trying to move towards a constructive engagement with IWRM in the South Asian context. This has been done by exploring a perspective based on comprehensive and multilayered sub-basin development, which blends formal and informal mechanisms of governance for promoting the livelihoods of people. The chapter further professes that there is, of course, no readymade blue print on this perspective; rather, these are ideas based on field experiences that are still in the formative stage. It is, therefore, imperative that the different perspectives are projected within the boundary concept of IWRM, and this chapter is precisely an attempt in that direction.

Gender and Water

In the past decades, the development planners in South Asia completely ignored the crucial roles that women played in the water sector. It was assumed that women were only concerned about water for domestic uses and that it was only men who were responsible for the productive uses of water. Consequently, this blinkered perception not only led to unsustainable development interventions, but also underestimated women's productive roles, thus adversely affecting gender equity. Fortunately, there has been a definite change in this attitude, and recent water policy documents do often make mention of the importance of women as managers of water. All this has come about because of several reasons and initiatives. Gender mainstreaming came into "international prominence as a strategy" during the late 1980s (Krishna 2008: 207). Following this, from the 1990s, gender mainstreaming gained currency in water sector planning too. As international organizations working in the water sector recognized the need for gender mainstreaming in water (GWA and UNDP 2006), various policy initiatives, training programs and activities were launched to bring greater visibility to the gender dimension of water resource management. As a result, strategies for integrating women in all areas of policy and programs, and at all levels – from the identification of needs to project planning, implementation, monitoring, and evaluation – became popular among the donors. Thus, the last decade has seen an increasing interest in the question of gender and water. In South Asia, this interest can be attributed to two main factors. First, women's movements and struggles have made a mark, and secondly, many funders have come to insist on its inclusion of of women in policy-formulation and implementation (Kulkarni 2009). The Gender and Water Alliance (GWA) report of 2003 titled "Gender Perspectives on Policies in the Water Sector" aptly states that (*a*) involving men and women in influential roles at all levels can hasten the achievement of sustainability in the management of scarce water resources, and (*b*) managing water in an integrated and sustainable way can contribute significantly to better gender equity, by improving the access of women and men to water and water-related services (Khosla et al. 2004).

Due to all these developments, the water policies and plans in the countries of South Asia have been following this approach of integrating gender into the water sector. For instance, the Bangladesh's National Water Policy of 1999 highlights women's participation in water sectors. The Indian National Water Policy 2002, which though does not use the term "gender" or "women", declares that a participatory approach is to be embraced by involving not only the various governmental agencies, but also the users and other stakeholders, in planning, development and management of water resources. The National Water Plan of Nepal, adopted in 2005, has posited gender participation as one of the principles of social development. Although Pakistan does not have a national water policy as yet, the agenda for new policy reforms in all the sectors undertaken since mid-1990s clearly aims at achieving women's empowerment and gender equity at all levels; more importantly, the National Drinking Water Policy of 2009 recognizes the key roles that women play in drinking water sector and calls for ensuring their participation in decision-making for the sector at all levels. Many of these policies and reforms, thus, clearly mark a paradigm shift in the water sector from a techno-centric, supply-driven model to a demand-driven and participatory model.

However, upon looking closely at these new policies and policy reforms, what is evident is that they follow a more functional or efficiency-oriented approach, wherein the focus is more on making women effective instruments of improving the management and conservation of water resources and, of course, the efficiency and success of programs and/or projects. What this has led to is the deliberate and blatant watering down of gender concerns to women's participation at the level of community and in the area of drinking water (and occasionally irrigation programs) only (Joshi et al. 2010). Thus, what we find are numerous water projects/programs that focus on poor rural women and their collectives to regenerate and effectively manage water, mostly for drinking purposes, and for irrigation, when a wave of Participatory Irrigation Management (PIM) programs swept the water sectors of some of the countries at the community level in which women's participation became a pre-requisite. The result has been that women's presence and participation has become more important than the process of selection/identification of women who can participate, or the degree and type of their participation. This view of involving women has persisted in the evolving policies and reforms, resulting in the situation in which fundamentally diverse policies and institutional arrangements have produced strangely similar outcomes reflecting a persisting gender inequity (Joshi 2011).

All this comes out very lucidly in the chapter by Sara Ahmed in Section Two, wherein the author, in tracing the water sector reforms in the state of Gujarat in western India, examines the implications of decentralization of policy-formulation and implementation for rural women's participation in the new community water institutions, by undertaking a case study of women's participation in drinking water supply programs and PIM projects. The author shows that despite a significant representation of women in *pani-samities* (water collectives), there are a number of factors that constrain their effective participation and raise questions about their expected role in water management. Furthermore, the responsibilities commonly ascribed to women – for instance, keeping the area around all village water infrastructures clean and ensuring that no one wastes water – reinforce gender stereotypes, as these are essentially an extension of their unpaid household work to the public arena. The chapter stresses that women's empowerment cannot be achieved by separating and isolating them from the complex social relations underlying their myriad and diverse relationships with water; the environment; and the larger socio-economic, political and cultural contexts, within which a gendered analysis of decentralized water governance is embedded.

Another glaring gap in the effort to mainstream gender in the water sector is that the planners still unequivocally place women in the domain of domestic water use, assuming that uses of water by women mainly lie in the domestic or non-market sphere (Zwarteveen 1997). This assumption has dangerous connotations leading to gender inequity, since it implies that "women do not need water as individuals or as farmers, as they will indirectly benefit from their husbands' rights and access to water" (Upadhyay 2005: 412). Thus, water for irrigation and the irrigation sub-sector remains a male domain. While exploring and analyzing the masculine nature of irrigation sector, Zwarteveen (2008) points out that "[t]he feminist project in irrigation to date has largely been a project of representation of women" and that this "idea of representing women and of 'making women visible' was and is, of course, based on a parallel assumption that men are visible and well represented" (ibid.: 111). However, such attempts in most cases are only a lip-service. The chapters by Pranita Bhushan Udas on Nepal and by Shaheen Ashraf Shah and Nazeer Ahmed Memon on Pakistan illustrate this point.

Bhushan's chapter maps the irrigation sector in Nepal where there has been a change from decentralized to centralized water management with the state intervention in 1920s. The author argues that although the agenda of gender water equity was included in water policies and plans in the 1990s, the content of water policy and its implementation have been limited to making women visi-ble in water committees. The chapter explores and explains the reasons for this limitation. The author argues that this inequity is linked to the unequal benefits that one receives from the construction of water infrastructure, and that the irrigation policy is silent on other issues of inequity such as disproportionality of labour contribution in the construction of an irrigation canal to the size of landholding. The chapter stresses on the need to address these issues. It also argues that addressing such issues that help women and men water users (from different socio-economic backgrounds) to access water resources in a fair manner also implies incorporating gender water equity concerns.

Shaheen Ashraf Shah and Nazeer Ahmed Memon's chapter is a case study of women and water reforms in the Sindh province of Pakistan, which reveals that women's participation is minimal in water management and water institutions are gender-blind in policy-making and implementation. Evaluating the level of women's representation in the reform program and its impact on overall water management in the province, the chapter shows that despite reforms having been carried out in the province, women have never been consulted during the conception, planning, design, or implementation of any PIM program although evidence shows that women do use irrigation water for both productive as well as domestic purposes. The chapter highlights the major problems associated with women's active participation in community/water users associations at the grassroots level, stressing that due to the lack of visible participation of women in water management boards, they enjoy less privilege or access to opportunities in irrigation reforms. The chapter further argues that for effective, efficient and equitable management of water resources, men and women should be equally involved in the consultation processes, and in the management and implementation of water-related services.

A crucial arena when it comes to the issue of gender is empowerment, i.e., empowerment of women. In the agenda of most of the new policies and reforms, it is assumed that women's involvement and participation in water programs (community programs) are automatically empowering. True, participation and empowerment are closely related, but mere participation does not automatically mean that power relations are challenged, a criteria for empowerment. Participation is a means to empowerment, but not synonymous with empowerment. In order for participation to promote and lead to empowerment, it needs to be more than a process of consultation over decisions already made elsewhere. The claim that women's empowerment is the ultimate objective of many development policies and programs, has led to a demand for indicators of empowerment. These indicators are needed to understand the extent to which women are already empowered, and to evaluate if such policies and programs have been effective in achieving their stated aims (Oxaal and Baden 1997). There are a variety of ways in which indicators of empowerment can be developed. Some of the indicators devised for measuring women's empowerment are the Gender Empowerment Measure (GEM) of the United Nations' Human Development Report (HDR) (UNDP 1995); indicators developed by the Canadian International Development Agency (CIDA) (CIDA 1997); and indicators developed by Hashemi et al. (1996) (cited in Oxaal and Baden 1997: 20), in their research on rural women's empowerment in Bangladesh in the context of credit programs in Bangladesh.

The chapter by Sayeda Asifa Ashrafi and Rezaur Rahman is in line with these researches. Taking a case study of the first phase of the Small Scale Water Resources Development Subprojects (SSWRDSP) in a village in Bangladesh, the authors have measured women's empowerment by taking certain indicators which they have categorized into Empowerment Index (EI) and Composite Empowerment Index (CEI). The study suggests that empowerment in most cases is directly related to the factors that have a significant contribution in local-level economy and, therefore, implications for poverty as well. Further, it puts forth the issue that empowerment and women's participation in water management have a functional relationship, as the more empowered women in the project areas also thought about their responsibilities regarding water management.

The Groundwater Challenge

Technological changes in agriculture and intensive use of groundwater led to a spurt in water exchange for irrigation in many areas of South Asia. Dense groundwater exchange markets developed in the early 1980s in the regions that were suitable for sinking deep tubewells, leading to debates over the nature of such markets and their way of functioning (Prakash 2005). The expansion of groundwater irrigation has transformed the rural economy, leading to significant increases in agricultural productivity and in incomes. Farmer investment in wells and pumps has driven this expansion on the demand side; however, the supply of cheap agricultural energy – usually electrical power – is a critical, though often overlooked, driver of the groundwater boom. One serious outcome of this boom in numerous regions around the world has been groundwater overdraft, whereby the quantum pumped out had exceeded the quantum replenished through aquifer recharge, water tables have declined and water quality has deteriorated (Scott and Shah 2004). Groundwater overdraft has, in turn, put many constraints on the access to and use of groundwater. Rapid growth in groundwater irrigation in South Asia has been the main driver of the agrarian boom, especially in the alluvial plains of India, Pakistan and Bangladesh; the agricultural use of groundwater in these two countries account for the bulk of such use of groundwater worldwide. On the upside, groundwater development has provided sustenance to agrarian economies and supported millions of rural livelihoods. On the downside, it has created chronic problems of resource depletion and quality deterioration (Shah et al. 2003). Apart from the many challenges, the issue of water–energy nexus and inequity in groundwater distribution are prominent in the literature.

In South Asia, little can be done to improve the groundwater economy without affecting the energy economy. The efforts to make the energy economy viable are frustrated by the farming community's often-violent opposition to the efforts to rationalize energy prices. As a result, the region's groundwater economy has boomed by bleeding the energy economy (Shah et al. 2004). In Section Three, the chapter by Aditi Mukherji, Bhaskar Das, Nabendu Majumdar, Bhaskar R. Sharma and Partha Sarathi Banerjee discusses the relationships between agricultural power supply and groundwater irrigation in the eastern Indian state of West Bengal, where metered tariff system was introduced in place of flat tariff system for agricultural pump-sets, as part of power sector reforms with the support of the World Bank. The authors study the consequences of this change in terms of its impact on the buyers and sellers in informal water markets that exists in the state: the metered tariff system has been found to benefit the sellers who doubled their profits because of lower electricity tariffs and higher selling rates. Buyers, however, have lost out because they have to pay 30–50 percent higher rates of water even after the introduction of metered tariff system. The chapter highlights that in the long run, there is a possibility of contraction of water markets resulting in buyers losing access to irrigation water at affordable prices. It suggests that the state government take measures such as subsidize more wells for poor buyers to ensure equitable water distribution.

The issue of inequity in groundwater access is a major cause of concern. A group of academicians have advocated for dense and competitive groundwater markets on the grounds of accessibility and the efficient management of the scarce resource, without unpacking the nuances of unequal social relationships, ecological and historical functions that have shaped and determined groundwater access and use. North Gujarat in India became a historic case for groundwater-based agrarian economy and triggered intense debate over management, distribution and use of groundwater in India (Kumar 2007). In the process of understanding the social and environmental change, the South Asian region also attracted much scholarly work on groundwater management that includes technological, economic and social studies on the changing ecological situation and its impact on the society at large. While some scholars have made a concerted effort to analyze and reflect on the possible role of groundwater development in helping to address the persistent problem of rural poverty in Bangladesh, Nepal and eastern India (Narayanamoorthy 2001; Vaidyanathan 1996, 1999; Kahnert and Levine 1993; Shah 1993), others have raised questions about the link between groundwater development and poverty alleviation on the grounds of persistent resource inequity and appropriation of a common-pool resource by a particular class of people for private gains (Moench 2004; Palmer-Jones 1994; Bhatia 1992; Prakash 2005). Taking this issue further, the chapter by S. Janakarajan and Marcus Moench reports that with the progressive decline in the water table farmers have resorted to the competitive deepening of the wells. This has resulted in the increased costs of well irrigation and further has resulted in a new inequity among the well-owners and between the well-owning and the non-well-owning farmers. Similarly, the urban water demands have increased tremendously for domestic and industrial purposes. While there has been an ever-raising demand for water, hardly has there been any effort to develop the infrastructure to treat the used water. This is dangerous and contributes to the pollution of the existing water stock, according to the authors. Therefore, water resources are under severe threat not only because of the ever-increasing competing demands (by various sectors), but also because of the diminishing quality caused by the discharge of untreated domestic sewage and industrial effluents. In the coastal regions, the problem gets compounded by seawater intrusion. The main objective of this chapter is to show how degradation of the groundwater resource base through over-extraction and pollution contributes to inequity, conflicts, competition and, above all, indebtedness and poverty.

Water Conflicts in South Asia

"Water conflict" is a term used to describe a conflict over access to water, which can happen between nations, states, regions, or different social groups. Conflicts can occur between agriculture and industry, and between fisherfolics and farmers. When it comes to reservoirs, flood control and irrigation demands are in conflict. Different regions have different kinds of conflicts, which need to be addressed differently. Conflicts arise when disagreements occur between people with varied interests. They can be categorized as belonging to two broad types: (*a*) inter-state transboundary, and (*b*) intra-state inter-sectoral. Inter-sectoral issues have to be seen at multiple levels of stakeholders where sharing of water is the root cause of a conflict (Malhotra 2010).

A wide range of water conflicts have appeared throughout history. Water has historically been a source of tension and a factor in many conflicts. However, water conflicts arise for several reasons, including territorial disputes, a fight for resources and for strategic advantage. The United Nations recognizes that water disputes result from opposing interests of water users, public or private. These conflicts occur over both freshwater and saltwater resources, and between international boundaries. However, they occur mostly over freshwater because freshwater resources are of utmost importance, although unequally distributed over the earth's surface. In the Middle East, lack of cost-effective technology for desalinization can deepen the crisis of water scarcity for all water users, whether corporate entities, governments, or individuals, leading to tension and possibly aggression. Also, as Hensel et al. (2006) point out:

> [S]everal edited volumes, such as Environmental Conflict (Diehl & Gleditsch, 2001) and Conflict and the Environment (Gleditsch, 1997a, 1997b), and special journal issues (*Journal of Peace Research*, 35(3)) have been dedicated to exploring the connection between conflict and the environment. While the empirical findings in this literature are varied, they show that the likelihood of militarised conflict is significantly increased by a number of environmental factors, including increased population growth (Choucri & North, 1975; Stalley, 2003; Tir & Diehl, 1998), contention over territory (Hensel, 2001; Huth, 1996; Vasquez, 1993), increased degradation of soil (Stalley, 2003) and land (Hauge & Ellingsen, 1998), and low freshwater availability (Hauge & Ellingsen, 1998) (2006: 387).

South Asia as a region has historically witnessed conflicts over water resources and countries therein have also evolved ways to avoid harsh relationships with neighbors. One such instance is the signing of the Indus Water Treaty (IWT) between India and Pakistan (1960) to ensure a balanced distribution of waters from Indus, whereby the right to use waters from its three tributaries, viz., Indus, Jhelum and Chenab, was granted to Pakistan, while that from its three other tributaries, viz., Ravi, Sutlej and Beas, was granted to India. The agreement apportioned approximately 80 percent of the annual runoff of the Indus system to Pakistan and the balance to India (Iyer 2001; Verghese 1997; Singh n.d.). Signing the treaty was a huge compromise for India, but helped to prevent some severe conflicts, leaving either side free to develop its own sources. "However, there is a widespread perception in Pakistan (which heavily relies on the Indus water system) that the Indian control of the Indus water-head can be misused to block water to Pakistan and devastate its economy," which, of course, India refutes and pledges its commitment to the IWT (Singh n.d.: 10). The conflict between India and Bangladesh arising out of the issue of sharing of the Ganga waters through Farakka barrage has been perpetually unresolved. Several meetings and consultations at different junctures have been held, but nothing conclusive has happened so far. India and Nepal also have inconclusive issues over sharing of river water owing to what both countries believe as unequal agreements that were made earlier. There are also anxieties about sharing the costs and benefits and the pricing of energy exports. However, the agreement on the building of the Mahakali cascade through the Pancheswar dam at the boundary of the two countries has ensured equal distribution of the costs and benefits.

Within India too, there have been several conflicts arising out of unequal distribution of water between states. A very pertinent example is that of the Cauvery waters, which, during British era, were controlled by the Madras Presidency, now in the state of Tamil Nadu, with an embargo imposed on the upper riparian in Mysore, now in the state of Karnataka. However, the lapse of this agreement generated rivalry between these states, long after independence. Karnataka is also involved in conflicts with Andhra Pradesh over the waters of Krishna river. The waters of the Rihand reservoir shared between states of Uttar Pradesh and Madhya Pradesh now have another competing user, the Singrauli coalfields which extract water for cooling purposes in the thermal power stations. Large dams too have been a source of conflict between groups which gain and groups which loose. The case of Sardar Sarovar dam on the Narmada river in Gujarat whose construction has led to the displacement of a large number of tribal communities in Madhya Pradesh has remained unresolved and sub-judice. Another case is that of conflict between Punjab and Haryana, where no conclusion has been reached regarding the unallocated waters of Beas and Sutlej although some waters had been transferred to Rajasthan for the Indira Gandhi Canal as per the IWT. The problem was further compounded when reorganization of the states of Haryana, Punjab and Himachal Pradesh took place leaving Himachal to resolve problems of resettlement and rehabilitation of people displaced by the Pong dam, a problem earlier faced by Punjab (Verghese 1997). About 15 percent of the aquifers are already in critical condition; the figure would possibly increase to 60 percent by 2030 (Briscoe 1993). The groundwater situation is more serious in the north-western states of Punjab, Haryana and Rajasthan, and the southern states of Tamil Nadu and Karnataka. In these states, farmers are using large amounts of groundwater since canals and rivers are slowly running dry. The heavily subsidized supply of electricity to farmers in these states has also encouraged them to extract groundwater, using electric pumps (Mahapatra 2005).

Section Four unravels the politics deeply embedded in these conflicts and in the access to power and resources through three different chapters. Changing geopolitical situations in several nations have seen the emergence of new conflicts and of newer dimensions to older conflicts, which may or may not get resolved. The deep roots of these conflicts, especially between uses and users of water, their nature, process and implications for policy-making and policy-implementation, have been explained by K. J. Joy and Suhas Paranjape in their chapter which discusses conflicts between different kinds of uses and users. Presenting some case studies of water conflicts all over India like Kaladeo National Park (agriculture vs ecosystem), Cauvery water dispute (peri-urban and urban), Gangal Canal (rural vs urban), they suggest the need for a change of mindsets in perceiving water and the way it is managed. Multi Stakeholder Platforms (MSPs) are being suggested as an institutional option for bringing different interest groups and stakeholders together for dialogue and exploring options which are more acceptable. Certain preconditions are required for MSPs to be successful. This is owing to the various views of people, and changing legal and policy terrains. Hence, the chapter concludes by underscoring the need for both legal and policy reforms, whereby legislations must look towards developing the necessary inclusive framework and sufficient space and institutional support for negotiation and renegotiation amongst different stakeholders around the critical issues confronting the water sector.

The chapter by Sanjukta Das specifically attempts to understand how the waters of the Hirakud dam near Sambalpur city in eastern India has been slowly claimed by several users over time and what are the underlying factors that leave the weaker groups to suffer more while the powerful gain. The dam was built primarily to control floods, provide irrigation to western Orissa to generate electricity, help in pisciculture and provide drinking water. But conflicts have slowly risen especially between the farmers, who complained about the violation of their user rights. Further, many areas of the tail-end of the canal have not been getting adequate water. There have been grassroots-level movements to protest against the allocation of water by the state government to the industries that have influenced the government in signing of memoranda of understanding (MoUs) in their favor. At such a juncture, tradable right to water use under the riparian principle is not possible and requires institutional reforms in Orissa. Another solution, the chapter suggests, could be to provide water rights to farmers and increase water use efficiency in agriculture by way of construction of water courses with financial support from industries, and divert a part of the excess water to industry.

The last chapter by Ramesh Ananda Vaidya and Madhav Bahadur Karki explores and examines how regional cooperation can be achieved between the countries of China, India, Nepal and Tibet, located in the Himalayan region, in order to not only ensure sustainability of the water bodies, but also protect them from the impact of climate change, another potential threat that may also cause severe hazards to these delicate ecosystems. The authors recommend that water be allocated judiciously and efficiently to its users by taking an integrated approach to address the multidimensional usage of water, as well as the multidisciplinary nature of resource management in Nepal. This approach includes a focus on integrated land and water management, which implies the integration of land-use planning and the practices in the upstream watersheds (together with its effects on quantity and quality of water available downstream). The chapter recognizes the scarce nature of water resources with competing uses and conflicting interests of the users thereby warranting the best possible resource management paradigm and institutions. Therefore, the authors argue for the need to take advantage of "externalities" while planning water management. The externalities reflected in the upstream–downstream linkages, whether for communities, districts and provinces within national frontiers, or across them.

Water in Changing Contexts: New Challenges

The politico-economic context in South Asia is fast changing and thereby resulting in new challenges for effective water resource management. Some major issues that have emerged are the rise of peri-urban areas, minimum environmental flow, recognition of local water knowledge and coordination between users and uses. Peri-urban areas have always been and continue to be neglected on the ground that they do not fall within the purview of either urban or rural, leaving these transitional areas in a complex web of interconnected issues, usually taken up by the rural administration which itself is not competent enough to handle some of the more complex issues. Moreover, development activities in the fringe areas of cities result in increasing mobility of production factors, such as capital, labour, technology and information. The infrastructural development of these fringe areas tends to make intensive demands on environmental resources and poses problems by eating into valuable natural habitats like wetlands and core forest areas, and causing loss of or damage to prime farmlands, resulting in the increase of impervious surface (Hasse and Lathrop 2003; OECD 1990).

A study of peri-urban dynamics by Narain and Nischal (2007) has found urban entrepreneurs taking a lead role as stakeholders and defining their rights over common properties like village ponds and lands that had been originally managed by the villagers through collective contributions and later sold annually by Gram Panchayats because of their inability or unwillingness to maintain them. The phenomenon of encroachment of forest land and agricultural land by brick and concrete-manufacturing industries at the expense of peri-urban agriculture and horticulture has been seen around Aligarh city and Hubi-Dharwad city in India. Farmers have been forced to shift their land from agricultural to non-agricultural uses; they are now contractors, laborers, drivers and even brick-kiln owners (Singh and Ashgher 2005 in McGregor et al. 2006; Nunan and Shindhe n.d.). These activities in the peri-urban areas modify the natural biogeochemical flows. As I. Douglas explains:

> Chemical transformations associated with manufacturing, food processing and urban building lead to the release of heavy metals, plant nutrients and organic compounds to the atmosphere and to soil and water bodies that in excessive concentrations may cause harm to living organisms. Many peri-urban residents cope with high local pollution levels every day of their lives (2006: 23).

The local people for lack of economic and social power are unable to negotiate for their rights to water, which had been once theirs. Also special economic zones (SEZs) consider these transitional areas to be the best locations for their activities and often get special pipelines laid to supply water, but, in the process, cause disastrous effects on the environment by reducing the rate of groundwater recharge, polluting the surface and groundwater reserves (SANDRP 2007). A mix of interests and perceptions within a peri-urban area often results in social compression or intensification manifested in conflict over natural resources and changing role of stakeholders (Dupont 2005). In his chapter in Section Five, Vishal Narain examines the implications of urbanization for water use and management practices in peri-urban Gurgaon and Faridabad, two of the fastest growing districts of the north-western Indian state of Haryana. He attempts to explain not only the environmental impact of this urbanization, but also the equity issues with regard to water for different social groups – issues often overlooked by the popular media which mostly concentrates on the compensation received by farmers due to land acquisition.

In the midst of various issues around water and discussions on how best to manage this precious resource, the importance of traditional knowledge around water has gained significance. Driven by the need to build infrastructure, traditional knowledge, although not documented, has been transmitted by word of mouth from one generation to the other, and respects the value of every drop of rain which gives life to people. The indigenous knowledge system also respects the agro-ecological zone diversity, and has developed a specific science, a relevant engineering and a technology appropriate to every geographical region. Such technologies range from capturing moisture from the sand the arid regions of Barmer in Rajasthan; to harvesting drinking water using *kuinyas* and *johads* (simple mud-built and concave-shaped barriers built across the slope to arrest rainwater run-off) in semi-arid zones around the Aravallis; and to harnessing the excess water (*ahar-pyne*), which is, in fact, a "flood water harvesting system," from the Ganga, driven by channels (*pynes*) deep inside the land, up to 30–40 km to fill tanks (*ahars*). This ensured a long-lasting retention of water throughout the year and a better distribution of silt (Singh 2009).

In higher altitudes, various conservation strategies to capture gravity water and store large amounts have been followed through generations, some of them being *kuhls*, *khatris*, *nawns* and *chhrudus* that have enabled sustainable use of surface and groundwater resources (Sharma and Kanwar 2009). Technologies like building a pond near a house, for bathing and washing and re-use of this water for kitchen gardening and roof catchments to collect run-off along hill sides are some examples of traditional knowledge put into practice even today in water-scarce zones of Nepal and also in Bangladesh and Sri Lanka (Sharma et al. 2009; Ullawishewa 1994; Chadwick et al. n.d.). Harvesting of spring water and its collection in a lined pond or tank for subsequent distribution among agricultural fields and fog collection centers in higher altitudes are also an example of the use of traditional knowledge for water management (Merz et al. 2003). In Sri Lanka, tank cascade systems of different types for different uses, associated with a variety of ecological and socio-economic subsystems also exhibit fine examples of traditional engineering techniques (Bandara n.d.). Bangladeshi farmers, too, have used several kinds of traditional knowledge to prepare themselves ahead of time for water-related disasters.

These water management structures were historically owned and maintained by the community and everyone had equal access to them. In modern times, redefining traditional discourse and relooking at the interventions made in the history of the community, which has triggered the development of an autonomous and endogenous system, in harmony with nature, can help sustain many of these hydraulic structures. This concept has been re-explored by Luisa Cortesi in her chapter on traditional water wisdom, with regard to the rejuvenation of dug wells in the five districts of Supaul, Saharsa, Khagaria, Madhubani and West Champaran in north Bihar. This activity has been taken up by the Megh Pyne Abhiyan in collaboration with a network of organizations working at the grassroots level.

♒

References

Agarwal, A., M. S. delos Angeles, R. Bhatia, I. Chéret, S. Davila-Poblete, M. Falkenmark, F. Gonzalez-Villarreal, T. Jonch-Clausen, M. AïtKadi, J. Kindler, J. Rees, P. Roberts, P. Rogers, M. Solanes and A. Wright. 2000. "Integrated Water Resources Management." Global Water Partnership/ Technical Advisory Committee (GWP/TAC) Background Paper 4, GWP, Stockholm.

Asaf, Hamed. 2010. "Integrated Water Resource Management," in Mohamed El-Ashry, Najib Saab and Bashar Zeitoon (eds), *Arab Water: Sustainable Management of a Scarce Resource*, pp. 91–106. Beirut and Lebanon: Arab Forum for Environment and Development (AFED).

Bandara, C. M. Madduma. n.d. "Village Tank Cascade Systems of Sri Lanka: A Traditional Technology of Water and Drought Management," http://drh.bosai.go.jp/Project/Phase2/1 Documents/8_ Proceeding/ 8_TIK6_P.pdf (accessed October 18, 2011).

Bandyopadhyay, Jayanta. 2009. *Water, Ecosystems and Society: A Confluence of Disciplines.* New Delhi: Sage.

Bhatia, B. 1992. "Lush Fields & Parched Throats: Political Economy of Groundwater in Gujarat." Working Paper, World Institute for Development Economics Research, Helsinki, Finland.

Biswas, Asit K. 2004. "Integrated Water Resources Management: A Reassessment", *Water International*, 29(2): 248–56.

Briscoe, John. 1993. "Incentives are the Key to Improving Water and Sanitation Services," *Water and Wastewater International*, 8(2): 28–36.

Canadian International Development Agency (CIDA). 1997. *Guide to Gender-Sensitive Indicators*. Canada: CIDA.

Cap-Net. 2005. "Tutorial on Basic Principles of Integrated Water Resources Management," http://www.archive.cap-net.org/iwrm_tutorial/main menu.htm (accessed September 15, 2013).

Chadwick, M., J. Soussan, D. Mallick and S. Alam. n.d. "Understanding Indigenous Knowledge: Its Role and Potential in Water Resource Management in Bangladesh," http://www.nrsp.org/database/documents/580.pdf (accessed March 24, 2011).

Douglas, I. 2006. "Peri-urban Ecosystems and Societies Transitional Zones and Contrasting Values," in D. McGregor, D. Simon and D. Thompson (eds), *The Peri-Urban Interface: Approaches to Sustainable Natural and Human Resource Use*, pp. 18–26. Earthscan: London.

Dupont, V. 2005. "Peri-urban Dynamics: Population, Habitat and Environment on the Peripheries of Large Indian Metropolises: An Introduction," in V. Dupont (ed.), *Peri-Urban Dynamics: Population, Habitat and Environment on the Peripheries of Large Indian Metropolises: A Review of Concepts and General Issues*, pp. 3–20. New Delhi: Centre de Sciences Humaines (CSH).

Gender and Water Alliance (GWA) and United Nations Development Programme (UNDP). 2006. *Resource Guide: Mainstreaming Gender in Water Management*. The Hague: GWA. Also available online at http://www.wsscc.org/sites/default/files/publications/gwa_resource_guide_mainstreaming_gender_in_water_management_2006.pdf (accessed March 7, 2013).

Hashemi, Syed, Sidney Ruth Schuler and Ann Riley. 1996. "Rural Credit Programs and Women's Empowerment in Bangladesh," *World Development*, 24(4): 635–53.

Hasse, J. E. and R. G. Lathrop. 2003. "Land Resource Impact Indicators of Urban Sprawl," *Applied Geography*, 23(2–3): 159–75.

Hensel, Paul R., Sara M. Mitchell and Thomas E. Sowers. 2006. "Conflict Management of Riparian Disputes," *Political Geography*, 25(4): 383–411. Also Available online at http://www .ias.ac.in/currsci /10sep2010/562.pdf) (accessed May 2, 2011).

Iyer, R. 2001. "Water-Related Conflicts: Factors, Aspects, Issues," in M. Mekenkamp, P. van Tongeren and H. van de Veen (eds), *Searching for Peace in Central and South Asia*, pp. 277–90. Boulder, CO: Lynne Rienner Publishers.

Joshi, Deepa. 2011. "Caste, Gender and the Rhetoric of Reform in India's Drinking Water Sector," *Economic and Political Weekly*, 46(18): 56–63.

Joshi, Deepa, Anjal Prakash and Chanda Gurung Goodrich. 2010. "The Rhetoric and Reality of Women's Roles in IWRM." Paper presented at the International Conference on Interdisciplinarity in Water Education: Challenges, Perspective and Policy Implications, organized by SaciWATERs, October 3–6, Kathmandu, Nepal.

Kahnert, Friedrich and Gilbert Levine (eds). 1993. *Groundwater Irrigation and the Rural Poor: Options for Development in the Gangetic Basin.* Washington DC: The World Bank.

Khosla, Prabha, van Wijk Christine, Verhagen Joep and James Viju. 2004." Gender and Water." Thematic Overview Paper, International Water and Sanitation Centre (IRC), The Hague, The Netherlands.

Krishna, Sumi. 2008. *Genderscapes: Revisioning Natural Resource Management.* New Delhi: Zubaan.

Kulkarni, Seema. 2009. *Situational Analysis of Women Water Professionals in South Asia.* Hyderabad, India: SaciWATERs and SOPPECOM.

Kumar, Dinesh M. 2007. *Groundwater Management in India: Physical, Institutional and Policy Alternatives.* New Delhi: Sage.

Mahapatra, R. 2005. "India Faces Water Conflicts," http://ia.rediff.com/news/2005/oct/06water.htm (accessed April 12, 2011).

Malhotra, Richa. 2010. "Water Wars and Resolution," *Current Science*, 99(5): 562.

McGregor, D., D. Simon and D. Thompson. 2006. "Contemporary Perspectives on the Peri-Urban Zones of Cities in Developing Countries," in D. McGregor, D. Simon and D. Thompson (eds), *The Peri-Urban Interface: Approaches to Sustainable Natural and Human Resource Use*, pp. 1–17. Earthscan: London.

Merz, Juerg, Gopal Nakarmi and Rolf Weingartner. 2003. "Water Scarcity in the Rural Watersheds of Nepal's Middle Mountains," *Mountain Research and Development*, 23(1): 14–18.

Moench, Marcus. 2004. "Groundwater: The Challenge of Monitoring and Management," in Peter H. Gleick and Nicholas L. Cain (eds), *The World Water 2004–2005*, pp. 78–100. Washington DC: Island Press.

Narain, Vishal and Shilpa Nischal. 2007. "The Peri-urban Interface in Shahpur Khurd and Karnera, India," *Environment and Urbanization*, 19(1): 261–73. Also available online at http://dspace.mdi.ac.in/dspace/bitstream/123456789/318/1/Peri- urban+interface+in+Shahpur+Khurd+and+Karnera.pdf (accessed September 22, 2011).

Narayanamoorthy, A. 2001. "Irrigation and Rural Poverty Nexus: A Statewise Analysis," *Indian Journal of Agricultural Economics*, 56(1): 40–56.

Nunan, Fiona and K. C. Shindhe. n.d. "Urbanisation Leading to Changing Land Use Trends," in Robert Brook, Sangeetha Purushothaman and

Chandrashekar Hunshal (eds), *Changing Frontiers: The Peri-urban Interface, Hubi-Dharwad, India*, http://www.bestpracticesfoundation.com/pdf/Peri_Urban_Book.pdf) (accessed September 12, 2011).

Organisation for Economic Co-operation and Development (OECD). 1990. *Environmental Policies for Cities in the 1990s*. Paris: OECD.

Oxaal, Zoë and Sally Baden. 1997. *Gender and Empowerment: Definitions, Approaches and Implications for Policy*. Report No. 40, Bridge Development – Gender. Brighton: Institute of Development Studies (IDS). Also available online at http://www.bridge.ids.ac.uk/reports/re40c.pdf (accessed March, 6, 2013).

Palmer-Jones, R. 1994. "Groundwater Markets in South Asia: A Discussion of Theory and Evidence," in Marcus Moench (ed.), *Selling Water: Conceptual and Policy Debates over Groundwater Markets in India*, pp. 11–47. Ahmedabad: Vikram Sarabhai Centre for Development Interaction (VIKSAT), Pacific Institute for Studies in Development, Environment, and Security, and Natural Heritage Institute.

Prakash, Anjal. 2005. *The Dark Zone: Groundwater Irrigation, Politics and Social Power in North Gujarat, India*. Hyderabad, India: Orient Longman.

Scott, Christopher A. and Tushaar Shah. 2004. "Groundwater Overdraft Reduction through Agricultural Energy Policy: Insights from India and Mexico," *Water Resources Development*, 20(2): 149–64.

Shah, T., C. Scott, A. Kishore and A. Sharma. 2004. *Energy–Irrigation Nexus in South Asia: Improving Groundwater Conservation and Power Sector Viability*. Research Report 70 (2nd revised edn). Colombo, Sri Lanka: International Water Management Institute (IWMI).

Shah, Tushaar, Aditi Deb Roy, Asad S. Qureshi and Jinxia Wang. 2003. "Sustaining Asia's Groundwater Boom: An Overview of Issues and Evidence," *Natural Resources Forum*, 27(2): 130–41.

Shah, Tushaar. 1993. *Groundwater Markets and Irrigation Development: Political Economy and Practical Policy*. New Delhi: Oxford University Press.

Sharma, Neetu and Promila Kanwar. 2009. "Indigenous Water Conservation Systems: A Rich Tradition of Rural Himachal Pradesh," *Indian Journal of Traditional Knowledge*, 8(4): 510–13.

Sharma, Subodh, Roshan Bajracharya and Bishal Situala. 2009. "Indigenous Technology Knowledge in Nepal: A Review," *Indian Journal of Traditional Knowledge*, 8(4): 569–76.

Singh, A. L. and M. L. Ashgher. 2005. "Impact of Brick Kilns on Landuse/Landcover Changes around Aligarh City, India," *Habitat International*, 29(3): 591–602.

Singh, Rajendra. 2009. "Indigenous Systems of Water Management and Their Modern Applications," August 10, http://www.indigenousportal.com/Traditional-Knowledge/Indigenous-systems-of-water-management-and-their-modern-applications.html (accessed October 4, 2011).

Singh, Richa. n.d. "Trans-boundary Water Politics and Conflicts in South Asia: Towards 'Water for Peace'," published by Democracy and Social Action, New Delhi. Available online at http://www.boell-india.org/downloads/water._Final.pdf (accessed September 5, 2011).

Solanes, M. and F. Gonzalez-Villarreal. 1999. "The Dublin Principles for Water as Reflected in a Comparative Assessment of Institutional and Legal Arrangements for Integrated Water Resources Management." Global Water Partnership-Technical Advisory Committee (GWP/TAC) Background Paper 3, Global Water Partnership/Swedish International Development Cooperation Agency, Stockholm.

South Asia Network on Dams, Rivers and People (SANDRP). 2007. "The SEZ threat to Water and Food Security," http://www.sandrp.in/otherissues/SEZ_threat_water_food_security_April2007/view?searchterm=ground water (accessed September 11, 2011).

Ullawishewa, Rohana. 1994. "Women's Indigenous Knowledge of Water Management in Sri Lanka," *Indigenous Knowledge and Development Monitor*, 2(3): 17–19.

United Nations Development Programme (UNDP). 1995. *Human Development Report 1995*. Oxford: Oxford University Press.

Upadhyay, Bhawana. 2005. "Gendered Livelihoods and Multiple Water Use in North Gujarat," *Agriculture and Human Values*, 22(4): 411–20.

Vaidyanathan, A. 1996. "Depletion of Groundwater: Some issues," *Indian Journal of Agricultural Economics*, 51 (1&2): 184–92.

———. 1999. *Water Resources Management: Institutions and Irrigation Development in India*. New Delhi: Oxford University Press.

Verghese, B. G. 1997. "Water Conflicts in South Asia," in "Water Conflict Part Two: Southern Asia and the United States," special issue, *Studies in Conflict and Terrorism*, 20: 185–94. Also available online at http://werzit.com/intel/ classes/amu/classes/lc514/LC514_Week_14_Water_Conflicts_in_South_Asia.pdf) (accessed September 22, 2011).

Zwarteveen, Margreet. 1997. "Water: From Basic Need to Commodity: A Discussion on Gender and Water Rights in the Context of Irrigation," *World Development*, 25(8): 1335–50.

———. 2008. "Men, Masculinities and Water Powers in Irrigation," *Water Alternatives*, 1(1): 111–30.

1

Towards Integrated Water Policies in South Asia

From Critique to Constructive Engagement

Amita Shah and *Anjal Prakash*

Integrated Water Resources Management (IWRM) is increasingly gaining currency in the contemporary discourse on water. The discourse, however, is influenced mainly by the two popular perspectives on water, viz., scarcity and crisis management. Therefore, focus of the discourse has shifted from development of water resources for its productive use and thereby poverty reduction, to demand management through pricing and centralized formal structures for governing water use and sectoral allocations using river basin as a basic unit. Originating as a response to ensure and further promote developmental as well as ecological functions performed by water, IWRM, as defined by the Global Water Forum (GWF), has brought into its fold a number of noble ideas such as coordinated efforts for water and land resources; maximization of welfare along with equity; and ecological sustainability (GWP 2000: 22). The definition though, acclaimed for pooling together a bunch of well-intended and uncontestable ideas, has been criticized for being ambiguous and imprecise and amenable to significant distortions leading to undermining of the very spirit of IWRM as a process-oriented, people-focused and sustainable development of land and water resources.[1]

Much of the criticism is based on the specific variant of IWRM, developed and implemented by the World Bank in some of the countries in Asia and Africa. This variant, recognized as the "mainstream" approach of IWRM (notwithstanding the scope for alternative interpretation of the very broad and least precise definition), consists of features like declaring water as state property; instituting water withdrawal permits; pricing of water for all uses except drinking and domestic use; and setting up river basin organization for deciding allocation of water.[2] The central thrust of the approach seems to be on centralized governance and pricing with nationalization of ownership and adoption of a larger unit for management such as river basin. The features that are important from the viewpoint of equity and sustainability are seen more as playing instrumental roles.

The contemporary discourse on IWRM in South Asia, therefore, is divided into three broad streams:[3] one, rejecting the IWRM approach for being politically maligned and using IWRM as a pretext for pushing the neoliberal agenda (Jairath 2008); second, questioning the suitability of IWRM in the context of developing countries predominated by informal water economies while accepting its relevance to the developing countries (Shah and Koppen 2006); and third, accepting the notion of IWRM as politically benign and thereby exploring alternative variants, particularly by trying to integrate formal and informal mechanisms that are multilayered and pluralistic in nature (Saravanan 2006).

Since most of the critiques have recognized IWRM as an ideal goal or an ideology worth exploring, it is imperative to ensure that the baby (i.e., the concept itself) is not be thrown out with the bath water. In this context, Peter Mollinga's suggestion for adopting IWRM as a "boundary concept" appears quite relevant. According to Mollinga (2006), the boundary concept would create a common ground and will allow different interest groups to interact more constructively.[4] While there may not be any monolithic perspective on what can be the constituent features of IWRM, a common ground can be built by drawing from the three important aspects of IWRM on which global consensus seems to have been achieved. These are: river basin unit; stakeholder involvement and privatization. These elements may serve as useful starting points for triggering processes of informed dialogue, which can then help to attain three important goals: (*a*) eco-system-based units for integrated land and water management with multilayered planning and governance; (*b*) people's participation leading to efficiency and accountability; and (*c*) blending of formal as well as informal markets and institutions.

A number of initiatives have already been undertaken in India and other developing countries by incorporating some or most of the features of IWRM noted earlier. Some examples of these initiatives are watershed development, sub-basin management, rainwater harvesting and institutionalization of groundwater markets. However, many of these have remained scattered, smaller in scale, and have generated impact in local/micro-settings as borne out by the experiences of a large number of watershed projects in the country.[5] Notwithstanding these limitations, these initiatives can work as building blocks for evolving a new perspective on integrated development of natural resources with centrality of water across different agro-ecological regions in the country. Unfortunately, such initiatives, though documented well, have remained outside the mainframe of the discourse on IWRM at national and international levels.

It is against this backdrop that this chapter tries to move towards a constructive engagement with IWRM in the South Asian context. This has been done by exploring a perspective based on comprehensive and multilayered sub-basin development, which blends formal and informal mechanisms of governance for promoting livelihood of the people. There is, of course, no readymade blueprint on this perspective; rather, there are ideas based on field experiences and still in the formative stage. It is, therefore, imperative that different perspectives are projected within the boundary concept of IWRM as mentioned earlier. This chapter is an attempt in this direction. It is divided in five sections. The next section critiques the "mainstream" IWRM. Sections Three and Four look at IWRM from the lens of watershed management and present an alternative middle path through multilayered approach using sub-river basin as a unit for planning and management. The last section presents a summary of the analyses.

Conceptualization of IWRM: Recapitulating the Debate

Global Water Partnership (GWP) defines IWRM as a process which promotes the coordinated development and management of water, land and related resources, in order to maximize the resultant economic and social welfare in an equitable manner without compromising the sustainability of vital ecosystem (GWP 2000). The definition, as noted earlier, neither has universal applicability, nor is binding under varying situations pertaining to resource endowment, stages of economic growth and socio-political structures across countries. In fact, the need of the hour is to explore multiple approaches with a view to arrive at a broad notion of IWRM, which suits the context-specific requirements as well as challenges of water resource management in each country.

The mainstream IWRM thinking looks at IWRM as one of the means to move away from the earlier sub-sector-based approach to a more holistic or integrated approach, which, *prima facie*, can address the emerging challenges of water resource management such as increasing scarcity (water stress), inter-sectoral conflicts, pollution, and lack of technical understanding on issues of water catchments. In this sense, IWRM is being viewed as a tool to mitigate the past abuse and ensure the future sustainability of water resources (Iyer 1994).

The main opponents of the "mainstream"[6] concept, however, treat IWRM as narrowly defined, underpinned by neoliberal principles, dominated by technical and managerial concerns, and informed by limited methodology and empirical data. They also point out the constraints in implementing IWRM such as (*a*) difficulties in the collection and use of social data corresponding to hydrological units; (*b*) limited technical capacity; (*c*) lack of integration between cultural aspects of water; and (*d*) non-congruence with the concept of decentralized water governance that is gaining ground in a number of developing economies. It is further alleged that using hydrological unit for implementation of an integrative planning may not be feasible, as it does not necessarily coincide with the political or administrative unit (Pangare et al. 2006).

Apparently, the contemporary debate on IWRM in India oscillates between two extreme positions. One stream accepts the broad concept but opposes it on the ground of operational feasibility, whereas the other stream opposes the very concept as being a part of the larger phenomenon of global imperialism. There is, however, a third stream, which treats IWRM as a phenomenon, independent from the debate on compliance and efficacy of the existing regulations, etc. In what follows, we recapitulate the important facets of the ongoing debate on IWRM in the country.

Shah and Koppen (2006) summarize the "IWRM package" that includes: (*a*) a national water policy for a cohesive normative framework; (*b*) water law and regulatory framework for coordinated action; (*c*) recognition of river basin as a unit for planning and management; (*d*) concept of water as an economic good to reflect its scarcity value; (*e*) creation of water rights; and (*f*) participatory water resource management and inclusion of women. However, critiquing IWRM on grounds of its non-feasibility and its irrelevance for developing economies, Shah and Koppen argue that IWRM is a process of transforming ungoverned water economies which do not have demand management systems in place to intensively manage water economies where direct and proactive demand management is a key feature. According to them, pathways to improved water governance through IWRM are treating water as an economic good, declaring water as a state property and creating property rights in water, and having appropriate water policies and laws in place. Further, IWRM also involves creating participatory structures and processes for water management at the river basin level. There are examples from developing countries where these changes have been effected. However, despite these changes, the water sectors of these countries are neither better governed nor are their people enjoying greater water-welfare.

According to Shah and Koppen, every country can transform their informal water economies to formal ones, but this transformation is mediated though iron laws of economic development. Shah and Koppen's model of the evolution of a water economy is based on the stage of its formalization which, in turn, depends upon the overall economic evolution of the economy. According to them, regardless of its water endowments, as a low-income economy climbs up the economic ladder, the organization of its water economy undergoes a transformation in tandem with societal transformation. They further argue that IWRM paradigm will not work in India because it is governed by a large informal water economy and hence does not have a formal class of intermediaries, i.e., water service providers for meaningful water demand management. According to them, water management goals can be addressed only through indirect policy instruments in South Asia at the moment to entice or compel private institutional arrangements and therefore it is better to focus on supply-side management, i.e., creation of better water infrastructure to have more efficient water supply. Further, as India urbanizes and Indian economy gets stronger, highly formalized segments will emerge especially in cities and hence direct demand management options will emerge, creating an ideal framework of IWRM to operate effectively.

Shah and Koppen's argument that India is not ripe for IWRM has been questioned on many grounds. Further, the argument can be extrapolated to similar economies of South Asia. Iyer questions the basic understanding and approval of the popular IWRM approach, which essentially is an "attempt to widen the hitherto dominant engineering perspective by including and internalizing environmental, social and human concerns, and enlarge the planning horizon from isolated projects to a larger hydrological framework such as river basin" (2006: 4623). While this is an improvement over the earlier framework, the approach continues to remain merely as a refinement of the engineering tradition and has an inbuilt tendency towards centralization and gigantism. He further points out that the issues of water as an economic good and pricing as a tool of regulation are derived from neoliberal economics and are part of the economic reform and structural adjustment programs, not necessarily linked to IWRM. Therefore, he asserts that pro-poor water policy should be independent of both IWRM and neoliberal "economic reform" programs. According to him, IWRM, neoliberal market fundamentalism and an advocacy of a national water policy and law are separate entities; they may overlap but are conceptually distinct. Hence, the question of whether to adopt IWRM or not should be discussed separately on its own merit. He also stresses the need for exploring effective ways for making water management systems work rather than shrugging one's shoulders and dismissively saying that policies, laws, regulation, and pricing do not or will not work in developing countries, such as India.[7]

Saravanan (2006), while critiquing Shah and Koppen's understanding of IWRM, argues that IWRM in its current package has been dismal worldwide. Citing examples from the US, Australia, South Africa and India, he shows that any economy is a mix of formal and informal mechanisms and both have their advantages and disadvantages. According to him, it is the combination of these that contributes towards integrated water resource management. The way forward implementing IWRM is to understand the process of integration of both formal and informal instruments of the water economy. In a complex adaptive economy, diverse actors ranging from international donor agencies to individuals evolve both formal and informal mechanisms to govern the water economy. Therefore, no one instrument of management is superior to other as each of these will have advantages and disadvantages. For instance, formal economies are better positioned to bring in macro-social and physical changes and values, but they are rigid, have high transaction costs and low pay-offs, and bring in commonly prevailing rules under a particular jurisdiction. Informal economies are better positioned to reflect social values that cut across administrative jurisdictions, have less transaction costs and high pay-off and are easily adaptable to growing uncertainty in water economies. Therefore, it is wrong to presume that it is only a formalized water economy that can be externally catalyzed, while informal economies are not "ripe" for IWRM. According to Saravanan, the answer lies in evolving adequate conceptual and analytical tools to understand the complex process of integration rather than succumb to ideological and normative presumptions.

The discourse on IWRM in South Asia has more or less come to a standstill, not making much headway from the highly contested positions taken by the proponents of large-scale, centralized mechanisms for water management and transfer on the one hand, and those advocating small-scale initiatives for augmenting and regenerating water resources, mainly ground water resources through micro-level watershed projects. These dichotomous views on water resource management, however, are entirely unwarranted in the light of the various initiatives, being undertaken at micro- and meso-levels. A more meaningful discourse may, thus, call for a constructive engagement that addresses the basic tenets of IWRM discussed earlier.

There is probably a middle path of adopting watershed-based approach, at sub-river basin level, for Integrated Management of Natural Resources (I-NRM). The single most important feature of this approach is that it seeks to simultaneously address the objectives of resource conservation, use-efficiency, livelihood sustainability and decentralized planning and management. There are, however, certain inherent constraints in the manner in which micro-level watershed projects are presently being implemented in the country. The major constraints pertain to issues like upstream–downstream conflicts, emphasis on in-situ management, scale disadvantage, reinforcement of present property rights regime, and individual stakeholders. Notwithstanding these limitations, Watershed Development Program (WDP) could perhaps, be treated as closest to the approach of a boundary concept of IWRM suggested by Mollinga (2006) though much of these interventions are too microscopic, without upward linkages to management at sub-basin or basin levels. How far can this help in evolving a context-specific approach to the implementation of IWRM in India? This question has been addressed in the next section.

IWRM and Watershed Development: Is There a Meeting Point?

In spite of being one of the most important policy initiatives in the recent past, WDPs in South Asia have remained isolated from the mainstream thinking on water resource management in the country. Strangely, one finds a complete disjuncture between watershed development and irrigation projects of various scales. In fact, the two are viewed as alternative approaches meant to serve different areas within each agro-climatic zone. The areas with potential for development of surface irrigation are expected to be outside the ambit of watershed projects, which, in turn, are expected to focus mainly on dryland areas with a special mandate on replenishing groundwater resources.[8]

It is plausible that IWRM, taking river basin as a unit of management, can create conditions for a well-synchronized approach wherein development of surface irrigation and recharging of groundwater through watershed development are integrated. In turn, this may also help in striking a balance between economic and ecological functions of water within a geo-hydrological unit. Unfortunately, a coordinated approach such as this has neither been clearly worked out nor been implemented on a larger scale. This is evidenced by the compartmentalized approaches adopted for sustainable management of natural resources (especially water), economic growth and poverty reduction, and water governance at different levels. This is a serious problem insofar as it perpetuates irrigation-centric, engineering-oriented, and bureaucracy-driven approach for water resource development with limited concern for equitable distribution and sustainability of water use. Some of these features appear glaringly in the recent Water Policy of 2002 in India (Bandyopadhyay 2006).

Breaking this status quo though difficult, is still feasible, if IWRM as a boundary concept is accepted for synchronizing the two major programs for water resource management, viz., irrigation and watershed development. This, essentially, would imply changing the composition of both, rather than merely building an administrative link between the two. In the crudest sense, this would imply pushing up micro-level watershed projects to a higher scale of, say, sub-basin level and, at the same time, introducing and/or refining important features of WDPs such as decentralization, multilayered planning and multi-stakeholder institutional arrangements for governance and market development.

In the following section, we present a broad outline of an alternative perspective that seeks to link watershed and irrigation development in the Indian context. The outline draws upon the wide ranging experiences of watershed development in different parts of the country.

Linking Watershed and Irrigation Development at Sub-basin Level: An Outline

There are several advantages of watershed development, which, at least at a conceptual level, make it more suitable for adapting to IWRM. These are: (*a*) interactionist view of natural resources; (*b*) emphasis on regeneration and conservation of water; (*c*) focus on water scarcity and use-efficiency; (*d*) direct link with livelihood promotion; and (*e*) decentralized management through local institutions. While all these make watershed development projects a good candidate for becoming the main vehicle for attaining the objectives of IWRM, there are certain inherent limitations in the form in which the watershed development projects are being presently operationalized in India. These include: (*a*) disadvantage of small scale that limits their potential to use resources and support livelihood; (*b*) excessive emphasis on in-situ conservation with limited scope for water transfer in times of scarcity; (*c*) neglect of upstream–downstream conflicts; (*d*) reinforcement of the existing property rights regime that leads to inequitable access and non-sustainable use; and (*e*) greater focus on individual benefits rather than societal/ecological benefits.

It is thus imperative that watershed development outgrows its present limited scope and scale so as to be able to attain the goals of IWRM on a larger scale. In the process, it may also trigger corresponding changes in the manner in which irrigation projects are designed and managed.

Bottom-up Approach

The proposed approach towards linking watershed development and irrigation development at the sub-basin level represents a bottom-up approach towards integrated planning and management of water and other natural resources. This implies mapping of resources and livelihood needs of the primary stakeholders at micro-watershed level as a starting point, which would provide base for a multilayered planning at the levels of milli-watershed and sub-basin. The multilayered planning would involve two-way flow of information, starting from the bottom-most level. At the same time, the bottom-up approach may imply assigning top priority to the needs of the primary stakeholders. Their needs, however, have to be defined within the ecological limits of sustainable resource use with inbuilt mechanism for attaining equity at least in the distribution of incremental water resources (Shah 2007).

The water requirements of primary stakeholders at micro- and macro-levels, however, should be negotiated through iterative processes in the light of the costs and benefits of different scenarios of sectoral allocations. This will help in not only arriving at socially contextualized and informed choices, but also evolving a rationale for pricing and differential rates of subsidies including cross-subsidization. This may even include zero pricing for the very poor. A decentralized approach may also help in creating incentives/disincentives for promoting right kinds of crop-mix, dissemination of information and inputs, or technology-mix suitable for the agro-ecological and socio-economic conditions specific to a particular region. On the other hand, relatively larger size of management unit may help developing markets and other related infrastructure for processing farm produce for value addition.

Need to Modify Institutional and Market-based Solutions

The iterative processes of negotiations and decision making noted earlier appear to be fairly complex. But the complexity is not a insurmountable problem as indicated by the local experiences of managing management tanks and other water bodies especially in the southern parts of India. This is not to suggest that these experiences can be replicated in the case of groundwater management where the resource is privately owned or controlled by landowners. Regulating the access to and use of ground water thus poses a much greater challenge. Development of water markets in such a case emerges as an effective solution given the property rights regime and the large magnitude of private operators in groundwater markets across the country.[9]

While these are important developments, and also amenable to modifications within the existing political economy framework, it is essential that traditional institutional arrangements (in the case of surface water) and newly emerging water markets (in the case of groundwater) are made more responsive to the larger concerns of productivity, sustainability and equity. The contemporary discourse on water management in India suggests economic pricing of water along with reforms in power sector to ensure dependable supply of water at a given price and promotion of value addition of the farm produce as an effective mechanism for addressing the concerns noted earlier. In this case, market rather than the state or local institutions become an effective regulator of access to and use of water.

The approach, though valid, may not work if the state has limited capacity to reform the power sector. Also, it may leave out of its preview, many poor farmers in remote areas who may find it difficult to enter the water markets by paying the economic price of water. It is likely that many of these farmers may be pushed into further distress if the present levels of subsidies and/or lapses in the collection of power charges are withdrawn. In this scenario, depending mainly on markets may hurt the interests of the poor. At the same time, it may not help in reducing depletion of groundwater since the poor farmers of a given region, given their limited land holdings and financial wherewithal for purchasing farm equipments or irrigation water, are not likely to draw lot more water than their resource-rich counterparts in the same region. Similarly, the traditional institutions governing tanks and other water bodies are also facing new challenges under the changing scenarios of resource augmentation and management in the upstream areas.

We do not intend to get into the details of these otherwise well-researched issues in this chapter. Nevertheless, it is plausible that both market as well as institution-led approaches noted earlier may gain additional leverage if these mechanisms work in a mutually reinforcing manner, constituting part of a larger and multilayered management of sub-basin.

Upstream–Downstream Conflicts and Intra-Sub-Basin Transfer

Adopting sub-basin as a unit may also facilitate monitoring of water balance in an upstream–downstream context. Currently, it is neither appropriate nor desirable to monitor such effects given the small size as well as scattered spatial distribution of micro-watersheds in most parts of the country.[10] On the other hand, information base on critical parameters of water and hydrology is far from desirable. What makes it even worse is the non-availability of existing database in public domain (Bandyopadhyay 2006). In the absence of such information, it is difficult to even explore alternative perspectives and mechanisms for operationalizing IWRM. In this context, a multilayered approach at sub-basin level may open up avenues for creating the requisite database, given the smaller scale and decentralized structure of governance as compared to the mainstream IWRM approach at the level of river basin.

A systematic analysis of water balance between upstream and downstream areas as well as inter-temporal contexts may pave way for the transfer of water within a sub-basin and, at the same time, create a base for assessing appropriate compensation and the institutional mechanisms for operationalizing IWRM.

Implications for Changing the Irrigation Projects

The perspective on upward movement of the present approach to watershed development may call for simultaneous changes in the way in which irrigation projects are presently designed and managed. One of the important features of such changes in the planning process is the adoption of a comprehensive approach towards water resource management, in tandem with watershed treatments for soil-moisture conservation and rainwater harvesting for recharging of groundwater. At present, planning of irrigation projects, especially minor and medium irrigation schemes, is undertaken independently of the watershed treatments, which, by and large, caters to the local needs at micro-watershed-village level.

In the changed scenario, the planning of irrigation projects and watershed treatments should merge in a manner that maximizes overall resource augmentation without compromising the needs of the primary stakeholders and sustainability of the ecology. A comprehensive planning for the sub-basin may be able to address the issue of checking soil erosion in the upstream areas, thereby enhancing the effective storage capacity of irrigation structures in the downstream areas. This may imply revival and integration of river valley projects. This, however, creates trade-offs between increased access to water across upstream and downstream areas, calling for well-negotiated choices and compensatory mechanisms that have been discussed earlier in India.

A recent policy initiative for promoting recharging of wells for enhancing the capacity of groundwater irrigation deserves special mention in this context. The initiative addresses one of the most crucial challenges facing water (irrigation) management, where groundwater is emerging as the single most important source of incremental irrigation capacity in the country (GoI 2006b).[11] The critical issue, however, is one of adopting a coordinated approach by taking up larger watersheds as units for planning and management for groundwater recharge. This is pertinent since groundwater recharge is very much at the center of the ongoing program for watershed development across large parts of the country.

Another important feature of the planning of irrigation projects is minimizing the role of large dams/reservoirs, especially for irrigation purposes. A recent World Bank study by Briscoe and Malik (2006) underscores the need for regulating groundwater use through a stronger presence of the state, which should not only own but also invest substantially in public goods such as sewer and wastewater treatment plants. More importantly, it advocates participatory aquifer development associations with rights to decide the pattern of water use along with corresponding responsibilities to maintain the resource. A strategy for groundwater development such as this may go well with the sub-basin-level approach discussed here. In turn, it may reduce the need for large dams, especially in ecologically fragile regions of the Himalayas.

Of course, much of the argument for building big dams in this region is grounded on their huge potential for generating hydropower. However, there is a need to re-assess their role in meeting the demand of water for irrigation and domestic use in the light of the potentially adverse environmental consequences of building them. It has been argued that location-specific mechanisms for conserving water and economizing its use such that it helps in regenerating the ecosystem would mitigate the risk of external shocks and may go a long way in supporting agriculture in large tracts of dryland regions in the country. Much of these dryland regions are characterized by hard rock and hence have little potential for groundwater development (Shah 2006).

To a large extent, renewed emphasis of the state on big dams and interlinking of rivers is part of the same ideology, which promotes centralized and market-oriented structure for basin-level water governance. This is reflected in the critical emphasis on pricing of water and electricity, to be preceded by assured increase in service delivery, most probably by the private operators, on a large scale. Obviously, there is little scope of safeguarding the interests of the poor farmers in remote areas, if the efficiency and sustainability is sought to be brought about mainly through markets and large private-sector players. Thus, a comprehensive, multilayered approach for integrating watershed and irrigation development projects may help in overcoming some of the damaging effects of large dams and predominance of market in water governance.

Where to Begin From?

Adopting watershed-based IWRM offers a bottom-up approach whose main features are sub-river-basin approach; planning and sectoral allocation through iterative processes between macro-, meso- and micro-levels; more of demand-based vs supply-based planning; greater scope for intra-basin transfer of water (including virtual water); larger units of planning providing greater space for negotiations among stakeholders (across sectors and upstream–downstream regimes); and multilayered institutions offering equal platform for negotiations. Markets may also assume special regulatory role in this framework. This may involve:

(*a*) Allocation of water by sectors and households would necessitate well-negotiated differential pricing structure (with possibility of zero price for the weaker sections).
(*b*) Built-in incentives for water-use efficiency (land use, crop-choice, technology).
(*c*) Compensatory mechanisms for resolving upstream–downstream conflicts over water.

(*d*) Greater role for community-based organizations (CBOs) in ensuring participation of people.
(*e*) Better linkages with input–output markets (scale advantage).

The question is where to begin in improving and upscaling WDPs? The answer lies in going beyond milli-watersheds; integrating WDPs with medium and minor irrigation schemes; strengthening WDP-based federations, if and where they exist; building capacity for negotiations; and bestowing legal status on multilayered institutions. Some of the pertinent questions, therefore, are: does this approach interface with the ongoing debate on IWRM? How can the stalemate in the debate over irrigation versus WDPs among experts, policy-makers and even parishioners resigning to the duality and misplaced emphasis on decentralization *per se* be broken? How do we evolve a holistic perspective on WRM and I-NRM irrespective of the contemporary debate on IWRM?

The most important challenge facing the proposed approach, is the institutional vacuum and non-performance of participatory institutions – in both watershed and irrigation management scenarios. Building multilayered structures of institutions and making them work beyond the scope or mandate of project interventions would require not only continuous financial support, but also large-scale interventions of the local government. If water is so critical for human existence, economic development and mitigation of future conflicts over it, it is essential to widen the base of the reform process, going beyond markets, rhetoric of participation and the existing structures of local governance.

In this context, there has been a strong plea for moving away from the "reductionist paradigm" of the earlier regime (Iyer 2007). Similarly, stressing the need for a holistic paradigm shift in water management in the country, Bandyopadhyay (2006) calls for water systems to be viewed as integrally linked hydrological cycles, delinking of linear relations of economic growth with greater water supply. Restructuring the institutional framework with ecological perspective is critical to moving towards a holistic paradigm.[12]

Presently, there are no readymade models of institutional structures that can combine efficiency of markets and institutional mechanisms for conflict-resolution. The need, therefore, is to change the mindset on water resource management and bring water resource management into the center stage of social, economic and political arenas influencing the large, scattered and diverse agrarian economy in India.

Way Forward

The foregoing discussion highlights the fluid and inconclusive discourse on IWRM across the world over. While the concept has been criticized as being too advanced and unsuitable for the contemporary scenarios of water resources and their management in developing economies like India, others have objected to the the use of this concept as a veiled means of pushing the agenda of neoliberal economic development into large agrarian economies of developing countries. A case is also being made for adapting the existing approach for watershed development to the higher level of sub-basin, which is large enough to allow economies of scale and scope and yet not too big like a river basin, which may cut across fairly divergent agro-ecological and socio-economic scenarios and hence are difficult to regulate and manage.

The sub-river-basin approach, not fully developed, is yet to create its legitimate place in the current discourses on IWRM in India. This chapter has tried to define at least the broad contours of an alternative approach, which seems to have better roots in the diagnosis as well as the present policy scenario in India. The broad outline presented here, however, needs to be fleshed out by way of more future research into the feasibility of an alternative approach to IWRM and its relevance to the debate on IWRM, which is at present and at the conceptual level too centralized, uniform and market-oriented to address the concerns of small farmers, water-scarce areas and poor consumers of water.

The next step, therefore, is to engage with an alternative perspective on IWRM, which is more suitable, workable, welfare generating and, hence, desirable. In fact, it may not really matter whether an alternative perspective such as the one discussed in this chapter, can be called IWRM or not insofar as it has three basic features, i.e., it is process-oriented, people-focused, and sustainable. What is essential, therefore, is to initiate a constructive dialogue on what is being suggested as boundary concept, lest the otherwise promising idea may lose out for the want of a clearly defined and universally applicable notion of IWRM.

In this context, the National Water Policy may assume special significance insofar as it may set the tone for an informed debate on IWRM among various stakeholders in different parts of the country, facing different constraints and challenges.

An important step at this stage is to work towards a more integrated and holistic understanding of natural resources in general and water in particular. This would necessitate going beyond the departmental boundaries. This would necessitate a comprehensive understanding of ecological, socio-economic and political realities by adopting an interdisciplinary approach, in order to define lower as well as upper boundaries of what IWRM can achieve in medium and long run. This, in turn, may call for simultaneously reviewing the major policy documents and initiatives that deal with natural resources, growth and human welfare, ecological sustainability, local governance, and fiscal instruments in place.

The next stage, then, should be to initiate a public debate in the light of the received wisdom on the status of the natural resources and alternative perspectives on their management focusing on the three basic elements of an integrated approach noted earlier. It is high time that such a process of constructive engagement is triggered through local initiative so as to attain at least a national consensus on IWRM.

≈

Acknowledgements

The authors would like to acknowledge the speakers and participants of the Brainstorming Workshop on IWRM in India: Concepts and Practice organized by Water Aid India and Gujarat Institute of Development Research, Ahmedabad during February 26–27, 2007 where some of these concepts are deliberated upon. In particular, the support from Mr. Depinder S. Kapur, the then Country Representative of WaterAid is fondly acknowledged. However, the usual disclaimer applies for any misconception in presenting the issues. This article was earlier published as working paper by Gujarat Institute of Development Research, Ahmedabad, India.

Notes

1. After the GWP released its Background Paper No. 4 on IWRM in March 2000, Bradford Centre for International Development set up an Alternative Water Forum in 2003, which initiated a series of seminars challenging what was claimed to be a global water consensus by the GWP. For details, see www.bradford.ac.uk/acad/dppc/seminar/water/.
2. For details, see Shah et al. 2006; Shah and Koppen 2006.
3. Much of the debate on IWRM has taken place through articles published in *Economic and Political Weekly* during 2006. For details, see Iyer 2006; Shah and Koppen 2006; Saravanan 2006.
4. It is, however, to be noted that "whether the IWRM concept will operate as boundary concept does not depend on the words themselves, but whether concrete resource governance and management issues or conflicts require and force amore 'integrated' perspective and whether integrated groups involved in these processes will actively call upon the idea of an integrated approach" (Mollinga 2006: 33).
5. For a brief review of the impact of watershed development projects in India, see Joy et al. 2006; Shah 2004; Kerr 2002.
6. For details on "mainstream" concept of IWRM, see note 3.
7. In a similar vein, Kapur (2006) argues that the correlation between water poverty and economic poverty is not at the core of the issues in IWRM debate. Economically developed capitalist countries have a lower dependnece on agriculture as source of income and employement. Lower degree of economic poverty in these countries may not be necessarily due to lower water poverty in these countries. In that sense, the statistically significant correlation between the two may not be very meaningful.
8. "Water is a critical input for agriculture and this calls for more effective utilization of existing irrigation potential, expansion of irrigation where it is possible at an economic cost, flood forecasting and better water management in rainfed areas where assured irrigation is not possible. This is clearly an area where past policies have been inadequate" (GOI, 2006: 24).
9. For details, see Shah 1993; Prakash 2005.
10. A study presently undertaken by the Forum for Watershed Research and Policy Dialogue in Karnataka, Maharashtra and Madhya Pradesh, is seeking to understand, among other aspects, the impact on hydrology on the downstream areas of fairly well-treated milli-watersheds, one each in the three states. For details, see Joy et al. 2006.
11. The official estimates suggest that there has been a six-fold increase in the net area irrigated through wells: the area irrigated through wells in-creased from 6 million hectares (ha.) in 1950–51 to 35.3 million ha.

in 2003–04. Compared to this, the area under major and medium irrigation projects increased from 8.3 million ha. in 1950–51 to 15.1 million ha. in 2003–04 (GoI 2006).

12. A non-governmental organization (NGO), Tarun Bharat Sangh, which has been working on water conservation and the promotion and revival of traditional indigenous water harvesting systems in the state of Rajasthan, perceived a threat to its work from the National Water Policy of 2002. Representatives of Tarun Bharat Sangh had been involved in the drafting process of the national policy, but not all of their recommendations had been adopted, particularly those regarding private sector participation in the planning, development and management of water resources (http://www.righttowater.org.uk/code/advocacy_4.asp [accessed August 29, 2007]).

References

Bandyopadhyay, J. 2006. "Criteria for a Holistic Framework for Water Systems Management in India," in P. Mollinga, A. Dixit and K. Athukorala (eds), *Integrated Water Resources Management: Global Theory, Emerging Practice and Local Needs*, pp. 145–71. Water in South Asia, vol. 1. New Delhi: Sage.

Briscoe, J. and R. P. S. Malik. 2006. *India's Water Economy: Bracing for a Turbulent Future.* New Delhi: The World Bank and Oxford University Press.

Agarwal, A., M. S. delos Angeles, R. Bhatia, I. Chéret, S. Davila-Poblete, M. Falkenmark, F. Gonzalez-Villarreal, T. Jonch-Clausen, M. Aït Kadi, J. Kindler, J. Rees, P. Roberts, P. Rogers, M. Solanes and A. Wright. 2000. "Integrated Water Resource Management." GWP/TAC (Global Water Partnership/Technical Advisory Committee) Background Paper 4, GWP, Stockholm. Also available online at http://www.gwpforum.org/servlet/PSP?iNodeID=215&itemId=24 (accessed October 16, 2011).

Government of India (GoI). 2007. *Ground Water Management and Ownership, Report of the Expert Group.* New Delhi: Planning Commission, GoI.

———. 2006a. *Report of Sub-Committee on More Crop and Income Per Drop of Water.* New Delhi: Advisory Council on Artificial Recharge of Ground Water, Ministry of Water Resources, GoI. Available online at http://wrmin.nic.in/writereaddata/mainlinkFile/File722.pdf (accessed June 13, 2013).

———. 2006b. Towards Faster and More Economic Growth: An Approach to the 11th Five Year Plan. New Delhi: Planning Commission, GoI.

Iyer, Ramaswamy. 2007. "Towards a Just Displacement and Rehabilitation Policy," *Economic and Political Weekly*, 42(30): 3103–07.

Iyer, Ramaswamy. 2006. "Water Resource Management: Some Comments," *Economic and Political Weekly*, 41(43–44): 4623.

———. 1994. "Beyond Drainage Basin and IWRM: Towards a Transformation of Thinking on Water." Working Paper No. 1031, Center for Global, International and Regional Studies, University of California, Santa Cruz. Also available online at http://econpapers.repec.org/paper/cdlglinre/1031.htm (accessed September 30, 2007).

Jairath, J. 2008. "A Plea for Unpopular Wisdom" (review of *Towards Water Wisdom: Limits, Justice, Harmony: A Plea for Unpopular Wisdom* by Ramaswamy Iyer), *Economic and Political Weekly*, 43(7): 32–35.

Joy, K. J., A. Shah, S. Paranjape, S. Badigar and S. Lele. 2006. "Reorienting the Watershed Development Programme in India." Occasional Paper, Forum for Watershed Research and Policy Dialogue (For WaRD), Pune.

Kapur, D. S. 2006. "IWRM: Understanding the Resistance and Barriers," *Water Drops*, 3: 15–16.

Kerr, J. 2002. "Watershed Development, Environment Services, and Poverty Alleviation in India," *World Development*, 30(8): 1387–99.

Mollinga, P. P. 2006. "IWRM in South Asia: A Concept Looking for a Constituency," in P. P. Mollinga, A. Dixit and K. Athukorala (eds), *Integrated Water Resources Management: Global Theory, Emerging Practice, and Local Needs*. Water in South Asia, vol. 1, pp. 21–37. New Delhi: Sage.

Pangare, V., G. Pangare, V. Shah, B. R. Neupane and S. P. Rao. 2006. *Global Perspectives on Integrated Water Resources Management: A Resource Kit*. New Delhi: Academic Foundation, in association with the World Water Institute, Pune.

Prakash, A. 2005. *The Dark Zone: Groundwater, Irrigation, Politics and Social Power in North Gujarat*. New Delhi: Orient Longman.

Samuel, H. and G. Mohan (eds). 2004. *Participation: From Tyranny to Transformation? Exploring New Approaches to Participation in Development*. London: Zed Books.

Saravanan, V. S. 2006. "Integrated Water Resource Management: A Response," *Economic and Political Weekly*, 41(38): 4086–87.

Shah, A. 2007. "Equity Issues in Watershed Development: Note Submitted to the Working Group on Natural Resource Management Prepared for the XI Plan." Mimeo. New Delhi: Planning Commission, GoI.

———. 2004. "Watershed Development: Rapporteur's Report," *Indian Journal of Agricultural Economics*, 59(3): 664–77.

Shah, M. 2006. "The Bank Comes Full Circle," *Economic and Political Weekly*, 41(22): 2201–3.

Shah, T. 1993. *Water Markets and Irrigation Development: Political Economy and Practical Policy*. Bombay: Oxford University Press.

Shah, T. and B. van Koppen. 2006. "Is India Ripe for Integrated Water Resources Management: Fitting Water Policy to National Development Context," *Economic and Political Weekly*, 41(31): 3413–21.

Shah, T., I. Makin and R. Sakthivadivel. 2006. "Limits to Leapfrogging: Issues in Transposing Successful River Basin Management Institutions in the Developing World," in P. Mollinga, A. Dixit and K. Athukorala (eds), *Integrated Water Resources Management: Global Theory, Emerging Practice and Local Needs*, pp. 109–44. Water in South Asia, vol. 1. New Delhi: Sage.

PART II

Gender and Water

2

Challenging the Flow

Gendered Participation, Equity and Sustainability in Decentralized Water Governance in Gujarat

Sara Ahmed

The concept of *governance* provides a useful tool for rethinking gendered patterns of access to water and water management within the ambit of broader social relations and organizational structures. However, not only is theoretical analysis and debate on "good water governance" limited, it is also largely gender-blind. Notions of what constitutes good governance are typically rooted in normative principles of accountability and transparency or participation and partnership which, it is assumed, will eventually lead to "good outcomes" for water, people and the environment rather than look at how agency is exercised or negotiated through the more informal relations of everyday life, structured by power (Cleaver and Franks 2005).

Since the early 1990s, decentralization policies in the water sector in India have sought to frame a new role for the state: from a supply-driven *provider* of water services to one which is *facilitating* demand and *enabling* community management. Constitutional Amendments under which Panchayati Raj Institutions (PRIs) in India were restructured gave these local governing bodies the added responsibility of managing drinking water and sanitation. PRIs are now expected to be responsible for the choice of technology, the recovery of operating costs (through water user fees), and the maintenance of rural water supply and sanitation schemes through elected *pani samitis* (water committees) at the village level. At the core of this process of institutional restructuring is the realization that water is no longer a free good and that decentralized management is the only way to ensure sustainable, equitable and efficient water delivery. Based on principles of cost recovery from users, *pani samitis* are meant to address management inefficiencies through participatory planning and inclusive decision-making.

Women's participation is seen as integral to these new institutions not only because there is gender division of roles and responsibilities in relation to water collection, but also because there is an implicit assumption that women's involvement is empowering for them and will lead to more sustainable and gender-equitable outcomes (Singh et al. 2004; WEDO 2003). However, participation is not a panacea in itself, nor do participatory processes necessarily challenge internalized gender-based oppression or lead to self-efficacy (Coles and Wallace 2005). Empowering rural women through water management initiatives requires more than providing access to decision-making or technical training: it needs strategic reflection on the micro-discourses of power (Rowland 1997). This means questioning notions of a "hegemonic masculinity," understanding how gendered identities are continuously reconstituted by institutional change, and how women themselves perceive their transformative potential. Empowerment cannot be achieved by separating and isolating women from the complex social relations underlying their myriad and diverse relationships with water, the environment and the larger socio-economic, political and cultural contexts within which a gendered analysis of decentralized water governance is embedded (Ahmed 2005). Decentralization is a process, which needs to be *negotiated*, and the hard reality is that for poor and marginalized women, negotiation is being contested in an economic environment where policies of privatization, pricing and centralized, technocentric delivery systems dominate the political discourse on water management.

Drawing on research insights from a two-year project,[1] this chapter looks at the unfolding of water sector reforms in the state of Gujarat in western India. It seeks to interrogate two core objectives of decentralization: (*a*) does it provide space for democratic, gender-sensitive and inclusive participation, and (*b*) does it lead to more efficient, effective, equitable (gender-just) and sustainable water delivery and management?[2]

Decentralization and Water Sector Reforms in India: Bringing Back the State?

Crafting democratic and decentralized institutions for community water management to which concerns of equity, gender and sustainability are integral requires an understanding of the peculiar nature of water as a common pool, flow resource where conflict and competition at multiple levels, between upstream/downstream villages and between sectors, is not uncommon (Mollinga 2006). Equally, water plays an important role in the democratization of local-level resource governance, not only because it is at the heart of daily livelihood activities and decision-making can facilitate democratic change, but water also has the potential to generate revenue for local government in the form of tariffs levied on its use (Ribot 2003). Thus, the reform process in water sector has been parallel to the restructuring of the institutions of local governance, under the 73rd and 74th Constitutional Amendments (1993) giving local governing bodies the added responsibility of managing drinking water and sanitation.

Pani samitis, as sub-committees of the Panchayat, are expected to be responsible for participatory, village-level micro-planning of the use and management of water and land resources; collect community contributions amounting to 10 percent of the capital costs for village water supply works; and be fully responsible for meeting all Operations and Maintenance (O&M) costs (by way of collecting water user fees) and for ensuring technical and financial sustainability of water infrastructure. However, because state governments in many cases still control grants to the PRIs and have access to funding from bilateral and multilateral agencies as well as the central government, they continue to be the main providers (and decision-makers) of a minimum supply of water to rural areas (UNICEF 1997: 9).

Decentralization in the water sector presents us with an interesting mosaic of competing meanings and intentions: one concerned with facilitating participation through quotas for women on *pani samitis* (one-third representation) and through proportionate representation for marginalized communities, and the other concerned with more efficient and effective water management by bringing the institutions of decision-making closer to the people. Water for the state is both a public and a political good and decentralization, to a large extent, has been driven by political benevolence as well as administrative expediency, in a context where competition and conflict over access to water can cause governments to lose crucial votes. Thus, the state seeks to maintain its hegemony over this process by legitimizing forms of populist "control" with only the delegation of responsibilities through administrative restructuring or de-concentration, rather than the devolution of power.[3] Underlying or masking this process of state "accumulation and dispossession" (Harvey 2003) is the language of growth with equity and justice, which presents us with a somewhat hybrid agenda of neoliberal reform in the water sector. Perhaps, no state in India illustrates the challenge of decentralization better than Gujarat.

Gujarat: The Development Landscape

Since its formation in 1960, Gujarat has shown steady progress in economic growth and has been ranked 4th in India in terms of per capita income for the past two decades. However, agricultural growth has shown a steady decline in the state, pushing small farmers and farm laborers into spiraling poverty and forcing them to increasingly migrate to urban areas.

Growing industrialization in the post-reform period (1991 onwards) has seen massive capital investments in the infrastructural development of coastal regions, often at the cost of land and water resources acquired and exploited by industries through liberal government policies. Although Gujarat has the longest coastline in the country, coastal livelihoods are increasingly vulnerable to cyclones, storms and intrusion of saline water. Currently, 23 percent of the state's water needs are met from surface water sources and about 77 percent from groundwater sources, with agriculture accounting for the major share of water consumption. About 77.4 percent of the total drinking water requirement in the state is met from groundwater sources (Kumar and Talati 2000: 44). The level of groundwater depletion is alarming, and today the state falls into the "grey" category of groundwater exploitation where 65–80 percent of recharged water is extracted (Mahatma Gandhi Labour Institute 2004: 80). This has also led to a rapid deterioration in the quality of available groundwater: 15 percent of the villages are affected by excessive fluoride content in groundwater, 6 percent suffer from excessive salinity and another 5 percent from high levels of nitrates (ibid.: 81).

Not surprisingly, the state does not have an environmental policy, and the water policy, first drafted in 2002 on the basis of the National Water Policy (2002) and then redrafted and shared with civil society in 2004, is yet to be implemented. Apart from the diversity of hydro-geomorphology in Gujarat, the skewed pattern of development is characterized by the withdrawal of state from social sectors (health, education and basic services). Needless to say, the gender gap in this context is growing. The most notable indicator of gender inequality is the declining sex ratio in the state which fell from 934 in 1991 to 921 in 2001 as per the 2001 Census (GoI 2001), largely due to a drop by almost 50 points in the juvenile sex ratio (0–6 years) during this period. Female infanticide and cultural biases against the girl child and women, in terms of access to education, healthcare and basic resource entitlements, persist despite the presence of a number of NGOs working on women's empowerment and gender rights. Gender budget analysis of women-specific schemes shows that actual expenditure is far less than the budgeted estimates despite the ambitious Gender Equity policy (2006) chalking out details of the tasks/responsibilities of each department to ensure gender equality, as well as specific goals on gender equality to be met by 2010.[4] Unlike the state's water policy, which barely mentions women, the Gender Equity Policy calls for the equal participation of women at all stages of water resources planning and management?

Decentralization: Building Community Institutions for Water Management

While there is little historical documentation of the social organization of domestic water use and management in Gujarat, compared to irrigation (Hardiman 1998), traditionally communities were responsible for the management of their village water sources and distribution systems. Under the Panchayat system, community participation in and contribution to intra-village water supply was called for in the Gujarat Panchayat Act (1961). The schemes were usually based on locally available sources, particularly wells, and costs were not too high. However, community contributions at 25 percent of the capital costs were higher compared to those in the present model; O&M was the responsibility of the Panchayat or the beneficiary community (Sama and Khurana 2006: 50).

In 1979, the Gujarat Water Supply and Sewerage Board (GWSSB) was formed as an autonomous body to provide safe and adequate drinking water, largely through the development of local water sources in villages with a population of 500 or less and/or through rural pipeline schemes, funded both by the state government and bilateral agencies. Although in some villages, NGOs and women's groups were involved it was not till the implementation of the Santalpur Regional Water Supply Scheme (1987) to provide drinking water to 72 villages in drought-prone Banaskantha district that organizations, such as the Self Employed Women's Association (SEWA), were more officially involved (Ahmed 2005, 2002).

In April 1995, a government resolution (GR) was passed and later revised, calling for the formation of a *pani samiti* in every Panchayat where a drinking water scheme had been implemented. According to the revised GR (2002), each *pani samiti* should have about 10–12 members, one third of whom must be women, and marginalized communities should be proportionately represented, ward wise if necessary. Each *pani samiti* must have at least 4–5 members from the Panchayat including the deputy *sarpanch* and other interested individuals (e.g., teacher, local health worker, etc.), who are willing to work voluntarily. The chairperson of the *pani samiti* is to be elected by members and the term of the *pani samiti* is two years, though the term can be extended to coincide with the term of the Panchayat if necessary. While the *sarpanch* is not constitutionally mandated to be the president of *pani samiti*, in a number of cases both positions coincide. Each *pani samiti* works under the administrative control of the Block Development Officer (BDO) and the technical guidance of GWSSB (until the official handover of water infrastructure) and has to maintain minutes of meetings and other records.

In 2002, the re-elected BJP-led Gujarat government introduced the policy of *Samras*: a village Panchayat does not need to hold elections, but can select eligible members on consensus through a process of dialog. The rationale behind this policy is to prevent violence, which, according to the administration, can be fatal in some districts. The government provides a one-time grant of INR 65,000 to villages that have successfully "selected" their Panchayat members and, sometimes, to *pani samitis*. While discussing the merits or otherwise of this policy, is beyond the scope of this chapter, it does appear to raise critical questions of transparency and accountability in the development of local governance.

Decentralization: Expanding Sector Reforms in Gujarat

From the mid-1990s onwards, the project of decentralized water management began to significantly move ahead with the launch of the Ghogha project in the coastal, drought-prone district of Bhavnagar, building on the Santalpur experience of partnerships between the state, the civil society and the community-based organizations (CBOs).

The Ghogha Rural Water Supply and Sanitation Project (GRWSSP) was initiated in 1997 under the Indo-Dutch bilateral aid program to build local institutional capacities to set up and manage local piped water supply systems in 81 villages and one small town in Bhavnagar district. It was to be implemented by the GWSSB in partnership with three NGOs acting as Implementation Support Agencies (ISAs) whose task was to facilitate and capacitate *pani samitis*, with various information and communication initiatives, using a range of informal and formal media (street plays, songs, awareness posters). However, till 2002 progress in the project was slow as GWSSB was not interested in this new demand-responsive approach under which no traditional major external works were to be undertaken. Instead, GWSSB was promoting the large-scale Mahi pipeline (from the Sardar Sarovar or Narmada project) as a source, given that all their surveys, undertaken in two consecutive drought years (2000–01), showed limited groundwater potential in most villages.

In 2002, the Dutch government gave in to the pressures from GWSSB and the state government and accepted the incorporation and partial financing of the Mahi pipeline system, making the Ghogha project a dual source project (Reynders et al. 2004). While the ISAs played a critical role in facilitating *pani samitis* and women's participation, they largely accepted the project framework; ideas for change were rarely entertained, and coordination, clear communication channels, roles and responsibilities, as well as knowledge sharing, were poor. Participation was, in sum, largely instrumentalist, "only to sell the project, not to adjust it to the needs of the target villages," (Grijpstra 2006: 201) making it very difficult to not only redirect the project, but also for villagers to take ownership. Not surprisingly, many villagers prefer to use traditional water sources, augmented by the water resource management measures, as externally based water supplies (through pipelines) are not available regularly or at convenient times during the day.

It is within this approach to participation that Water and Sanitation Management Organisation (WASMO) was formed in 2002 as a *learning organization* to "promote new mind-sets capable of a genuine paradigm shift in the sector." (GSDWICL 2000: 81, abridged). Communication skills and human resource development opportunities within WASMO were critical. In April 2003, WASMO initiated the Earthquake Reconstruction and Rehabilitation (ERR) project to transfer the learning from Ghogha to the reconstruction of water works in 1,255 earthquake-affected villages in four districts with the support of 30 ISAs or NGOs. Unlike the case in the Ghogha project, water resource management – such as recharging of groundwater, soil and water conservation – were made integral to the project design. Although added focus was on sanitation, achieving targets in this respect are still constrained for a variety of social, cultural and institutional reasons.[5]

WASMO has formed core teams of engineering and social sector specialists to support Community Management Support Units (CMSUs) at the district level, an act which is the closest public interface between WASMO and rural water users. Although it has a relatively hierarchical structure closely linked to GWSSB, WASMO claims that it encourages innovation and provides an enabling environment, seeking to empower communities to manage their community water systems (WASMO 2008). "Good governance," according to the WASMO Project Director, "suits everyone. Our mission is honesty as purpose. Gujarat has been a state characterized by scarcity so decentralisation was important – and also a focus on water quality issues – we have created a demand for safe water. Political need and commitment at all levels is visible" (interview, Gandhinagar, June 2008). While women's empowerment, as distinct from women's participation, was not an objective for WASMO, it is clear that women's participation in *pani samiti*s has been behind much of the acclaimed success of water sector reforms in the state. Thus, the Project Director adds, "Women have the *right to ask* for (safe) water at their doorstep: WASMO hopes that by 2010, 75 percent of the villages in Gujarat and women in these villages will have access to safe water at their doorstep" (ibid.). In reality however, this will remain a "pipe-dream" for years to come, as there are a number of technical and institutional problems with the water infrastructure provided, affecting the regularity, quantity and quality of water available.

The Research Context: Women and Access to Water in Rural Gujarat[6]

Thirteen villages, spread over six diverse agro-ecological districts were *purposively* selected with NGO partners to illustrate best practices, both of state-led decentralization[7] and community-led water management.[8] Although the selected villages have different development trajectories, they all have *pani samiti*s being facilitated by NGOs, largely following the same principles of demand-responsive water management with community contributions.

The villages range in size from 59 households to more than 2,000 and have basic infrastructure (primary schools, transport and communication systems, Panchayats) but largely lack sanitation facilities and primary health centers. Literacy rates, particularly female literacy rate, are low and sex ratio, though near the state average, varies considerably even in the same district, e.g., from 798 in Mokarshivangh village to 1,023 in Chadura village, both in Kutch. Prior to the development of community water supply systems, women in all the villages spent considerable time and energy in the daily drudgery of collecting water. Women recounted waking up at 4 a.m. in the summer months to find water in their *virda*s (hand-dug shallow wells in river beds), or depending on the Panchayat to call for water tankers, which came irregularly and never had sufficient water, leading to frequent conflicts (Barot 1997). Sometimes, men would go with their bullock carts to neighboring villages which had local water sources to fetch water. Caste, class and age are all determinants of who collects water: those who own land have farm wells which they (and sometimes their neighbors, kinsfolk, depending on reciprocity) can access for domestic use; women from upper castes, e.g., Darbars, do not go out to fetch water because of their strict observance of *purdah* (female seclusion) while older women have also largely relinquished this task to daughters and daughter-in-laws. Interestingly, it is this category of older women, who are represented on the *pani samiti*s, as discussed later in this chapter.

Caste discrimination persists in some villages, particularly those in Surendranagar district (Prajapati 2005). As a result, older *dalit* women in Navagam village generally do not go to fetch water, preferring to send their children. Of the eight Cluster Storage Tanks (CSTs) built by AKRSP(I) in the village, only three had water at the time of our field visit, compelling *dalit* women to either walk to a well beyond the village or try to use the CSTs in the upper-caste hamlets. In Ghogha, a Muslim-dominated town, the continuing poor and irregular water delivery was ascribed to the discriminatory policies of the BJP state government, where even bureaucrats and water engineers were not perceived to be "neutral." NGOs try to resolve some of these conflicts through multi-stakeholder dialogue and shared learning processes, but they are powerless in the face of larger networks of socially and historically embedded power.

Who Are the Women on Pani Samitis*?*

Most of the *pani samiti*s in our sample villages had been functioning for 3–5 years and some, such as the ones covered under the Ghogha project, were already set for a second term of elections. Of the 160 members across the 13 *pani samiti*s, just over half (53 percent) are women, though the range varies from a low of 25 percent women to 100 percent in the two tribal villages covered. Our sample of 68 female *pani samiti* members comprises 80 percent of the total number of women on *pani samitis* in these 13 villages. The majority of women members are married (82 percent), fall in the age group of 30–45 years (75 percent) and have not had any formal schooling (66 percent). Married and older women have a higher degree of mobility in the social domain and more time at hand as they are past their child-bearing and nurturing years. Interestingly, the lack of education does not act as a deterrent to their selection as members of to *pan samitis*, though they tend to regard their limited formal knowledge as disempowering.

Data on annual income levels were used as an estimate for poverty and the participation of poor in governance. However, this data is an underestimate and a more detailed analysis would require looking at consumption expenditure and accounting for inflation, tasks that we did not have the time or resources for. Looking at this "crude" data, 37 percent of the women on the *pani samitis* come from households that have reported income levels below US$ 0.40/day to just over a US$ 1/day. In contrast, 20 percent of the sample includes women coming from higher-income groups, while the majority are women belonging to the median income category. This suggests that although the *pani samitis* are dominated by women from relatively better-off socio-economic categories, there are still considerable spaces for the poor to participate in, or at least be represented in. But many poor women maintain that they do not have time to attend meetings because they are mostly engaged in daily wage labor.

In terms of religion, the *pani samitis* are dominated by Hindu women; most of the Muslim women in our sample come from the *pani samitis* in Kutch where the Muslim population has traditionally enjoyed better social mobility and participation in the public domain than that in the rest of the state. Looking at the representation of marginalized communities – at least 16 percent of the women on the *pani samitis* come from Scheduled Castes (SCs) and 26 percent from Scheduled Tribes (STs) – though the figure of ST women is only representative of a couple of districts in our sample, namely the eastern district of Dahod and the southern district of Surat.

The majority of female *pani samiti* members (90 percent) do some paid work, either as agricultural or daily laborers in the flourishing, but poorly paying, diamond polishing business. Some are involved in the delivery of village social welfare services, such as primary healthcare, child-care (in *anganwadis*), mid-day meal service and teaching at the primary level, while others run petty businesses like tailoring or grocery shops. Women who are not able to work in the public domain are typically from upper-caste households where there are restrictions on their mobility. Most female members of *pani samitis* have some prior experience in political or social mobilization either as members of Panchayats, Self Help Groups (SHGs),[9] or other village institutions, such as dairy cooperatives; disaster mitigation committees; and watershed, forest management or education committees. Several of these women have held leadership positions in these community institutions, but only three of them in our sample currently hold official positions in their *pani samitis*, as president, vice-president and secretary – all of them from villages in Kutch where the NGO partner Kutch Mahila Vikas Sangathan (KMVS) has made significant efforts to involve women's participation in local governance.

Prior institutional experience is the most significant factor in determining *which* women are selected to the *pani samitis*. The second most important factor is availability of time for them to attend meetings and fulfil responsibilities as a *pani samiti* member as well as go for training programs and exposure visits. Other criteria, such as being articulate, willingness and capacity to work effectively, demonstration of leadership, including conflict resolution skills, and mobility, are also important determinants. Contrary to perceptions that women are nominated by their husbands or other influential men in the village, most of the women on the *pani samitis* had their names put forward by members of the community institutions they were part of, or sometimes the facilitating NGO: nearly 82 percent of the women were finally selected at village *gram sabhas*, attended by both enfranchised men and women.

Gender and Participation: Empowering Women or Ensuring Project Efficiency?

Participation is one of the most maligned concepts in development policy and practice with a diversity of definitions (Oakley et al. 1991) or typologies (Agarwal 2001). These typically range from *participation as means* – the instrumentalist, rational objective of efficiency underlying most project or programmatic approaches to working with communities assumed to be undifferentiated, to *participation as an end* – where the process of participation is seen to be empowering and transformative in itself (Cleaver 1999). Despite the significant representation of women on *pani samitis*, there are a number of factors that constrain their effective participation and raise questions about their expected role in water management.

The most visible level of formal institutional participation is meetings: 68 percent of women members define this as their primary responsibility. However, most *pani samitis*, particularly those in the villages where the water works have been completed, rarely have regular, if any, meetings and even when they do have meetings – at times that are appropriate, informed to most members and largely been selected in consultation with women – many women (20 percent of the sample) are not able to attend. Apart from being preoccupied with work, or finding meeting times and location inappropriate, many women who did not attend felt that there was little support from others on the *pani samiti*: male members dominated the discussions, and prevailing socio-cultural norms meant they could hardly express an opinion. So, they stopped attending meetings, as they soon found them to be a waste of time.

However, some women use other strategies for articulating their concerns: sometimes key issues are discussed in women-only meetings, such as the SHGs, and women appoint a "representative," typically an older, experienced woman to speak on their behalf. But given the intersecting identities of caste and faith, a so-called collective women's voice or gendered priorities, may be consciously or not, *excluding*. Although there are no direct examples, *dalit* women who are also often the poorest, claimed that they found it more difficult to speak in *pani samiti* meetings or even to participate in some of the women's groups in their village, like the SHGs.

Upper-caste women face gender discrimination rooted in patriarchy: although they have been nominated to *pani samitis*, often by their husbands, they are "prevented" from attending meetings. "I let my wife be nominated for *sarpanch* (head of the village Panchayat), but I attend meetings," explained Shakti Singh (*pani samiti* member, Jasapar village, March 2007, WASMO-Ghogha project). "At least, I have made her a *sarpanch*, even if only in name; in my grandfather's time that would not have been possible. Maybe our children will have women *sarpanchs* who will also work, but that will only happen after everyone is educated," he added.

Apart from attending meetings, other responsibilities women claimed they had were: monitoring the construction of village water works, which involved ensuring the quality of raw materials, maintaining the attendance register of laborers, paying contractors and sometimes purchasing raw materials. Training in quality control for monitoring construction is given by WASMO. While articulation of their views and awareness can be seen as empowering for women members, the other responsibilities they cited as being commonly ascribed to them reinforce gender stereotypes and the instrumentality of participation. For example, tasks, such as keeping the area around all village water infrastructures clean and ensuring that no one wastes water, are an extension of their unpaid household work to the public arena, privileging women as natural environmental caretakers (Rocheleau et al. 1996). In addition, women have also been assigned the responsibility to collect water charges as they are considered better persuaders, particularly in the case of households which are either too poor to pay or are reluctant to do so, as they are not direct beneficiaries of some of the community schemes (e.g., if they have their own farm wells). Women claim that this almost amounts to vesting in them the responsibility for the financial sustainability of *pani samiti*: "so, if money does not come in, i.e., if people do not pay the water charges on time it is seen as women's fault" (*pani samiti* member, in FGD in Mithivirdi village, Ghogha-WASMO, January 2008). Both roles are embedded in perceptions of rural women as innately altruistic and more "accustomed to voluntary work" (World Bank's Water Resources Management Strategy 1993, cited in Green et al. 1998: 264), a perception which is common across many water projects couched in the language of "reform." For example, in her study of a drinking water project in northern Rajasthan, jointly funded by the state government and a large German development bank, under the aegis of public–private partnerships, O'Reilly (2008) argues that there is an unquestioned relationship between women and water underlying women's participation in the project:

> Women's groups must first and foremost serve the purpose of making the water supply system sustainable in the long run, i.e. women must be mobilised to take responsibility for the water management of their village. The health and hygiene education objective and the empowerment and self-help objective are important but should be subordinate to this overriding goal (Project Social Side report, cited in O'Reilly 2008: 199).

It is this goal – sustainable water supply systems and services – which is at the heart of water sector reforms, driving an instrumentalist rather than an enabling and empowering decentralization agenda.

It is too early to judge whether women's participation has led to their empowerment – certainly, access to improved water supplies (and, in some cases, sanitation), despite being irregular and at a cost, has largely had positive benefits on women's time, health, hygiene and their ability to send their daughters to school.[10] Their knowledge about the rationale for forming *pani samitis* and making decisions regarding the water distribution system and collective setting of tariffs was generally high, even though they may not have always agreed with some decisions and only 51 percent claimed that they had directly contributed to decision-making on the location of village water infrastructure. Most of the women – 72 percent of our sample – felt that their families supported their participation and that *pani samiti* meetings, largely because they were so infrequent, were not encroaching on their time. However, support from men, particularly in terms of sharing in household chores is still limited, and most of the costs of women's participation continue to be borne by their daughters and daughters-in-law, in effect the *actual* water collectors. In the Jasapara village dominated by the Darbar caste, men stated that women's participation should not be at the cost of their primary responsibilities: "We men have to wait for our meals if women go to meetings!" retorted Mohanbhai (*pani samiti* member, FGD, January 2007). Across all villages, most women continue to collect water and assume responsibility for health and hygiene of children.[11]

Engendering Water Governance?

Our findings in Gujarat have illustrated that decentralization, like democracy, is a process and not a goal. Building "technical" capacity to develop, manage and maintain village infrastructure for water supply is not the same as building the capacity of *pani samitis* to constructively engage in the process of decentralization as citizens and subjects rather than "objects" of participatory water planning. Political articulation – the degree to which communities of water users represented through *pani samitis* can influence water policy – is determined by the institutional architecture which facilitates interaction between civil society, the state and citizens in a wider democratic context (Chhatre 2007). The impact of water sector reforms and the space for rural women to participate in decentralized institutions, as well as for their empowerment, has to be looked at within this larger framework. In other words, it means looking into whether decentralization is contributing to more efficient, effective, equitable and sustainable water management and governance.

In terms of efficiency, facilitating demand-responsive water management has required tremendous financial and human resources as well as time. Little was achieved in the first five years of the Ghogha project; it was only with the formation of WASMO and the powers and resources that it had at its disposal that we begin to see serious efforts at developing *pani samitis* and facilitating partnerships between them, NGOs and the state. While we were not able to do a financial analysis of costs overrun or alternative costs (e.g., cost of a conventional rural water supply scheme) it is clear that without the collaboration of NGOs with substantial experience of community-led water management, innovative approaches as well as gender mainstreaming, this program would not have been able to meet its intended goals. That said, however, problems in the delivery of water, in access (physical and financial terms), water availability and quality persist in many villages raising questions about the effectiveness of sector reforms. These concerns were largely expressed by non-members, many of whom did not know what was happening on the *pani samitis* or who was representing their interests.

Perhaps, the most surprising finding was the internalizing of the discourse on sector reform – nearly 90 percent of the women in our sample agreed that all users should pay for water, and no one felt, not even the poorest, that people could not afford to pay for water. However, opinions varied regarding *which* people should pay *how much* and whether a crude "willingness-to-pay" argument could be extended to paying more for better quality water and water delivery services. Only 49 percent of the women were ready to pay more for water if it was available at their doorsteps (private connections), while 24 percent of the sample thought that the very poor should be exempted from paying (or be made to pay less than others) and that there should be some flexibility in payments, particularly for those who depend on daily wage labor. In some villages, *pani samitis* have made concessions for very poor households, for example, in Hajipir village, Kutch, the *pani samiti* built a separate storage tank for the 10–15 households of Koli Patels (considered a backward caste) at no cost (i.e., the 10 percent community contribution was waived), but asked them to pay the applicable water charges.

The biggest challenge is collecting the community contributions and water tariffs – a task that women already felt had been unfairly allocated to them. In many cases, there are examples of contractors, the *sarpanch* or better-off families providing the bulk of the contribution, a minimum condition before GWSSB or WASMO would sanction the construction of the village water works. But once schemes were designed and handed over, numerous problems of water delivery, lack of power for pumping, or simply inadequate capacity and pressure to reach tail-enders in the distribution system surfaced and some people were reluctant to pay water tariffs. In Ghogha town, people were even willing to pay more to get the occasional water tanker, as they felt the scheme had not delivered on its promises: "Unless you pay for water, you should not complain about poor services," exclaimed one of the women *pani samiti* members in an FGD, trying hard to get the 10 households in her hamlet to contribute. In some villages, the external Mahi water supply was cut because of non-payment.

Inequity in access to and availability of water is not only linked to the design or functioning of the scheme, but, more importantly, to the absence of norms regarding water entitlements, i.e., how much water and of what quality is a household or an individual entitled to? Although the schemes have been designed on a 40 liters per capita per day basis (the official norm for rural areas), there is no "institutional" design to ensure that when water is available, consumers would get their due share. In a few villages, the *pani samitis* had rules in place to prevent people from pumping water directly from pipelines or tanks, while in Dador village, Kutch, the *pani samiti* made a conscious decision not to provide private household connections, as in the absence of metering it would have been difficult to monitor water use. However, in most villages, rules about non-wastage of water are largely unwritten and therefore not always perceived as "formal" or enforceable. But rules on their own are meaningless, particularly when enforcers are largely female *pani samiti* members without legitimate power. Besides, a culture of change and developing values of civic or social responsibility is not a problem of rural India alone.

Questions of good governance are linked to transparency and a number of steps have been introduced to minimize corruption. For instance, all major decisions are taken at the *gram sabha* level, all accounts are audited, and *pani samitis* are empowered to decide on *who* should build their village water works. While it is difficult to measure corruption (Asthana 2004), WASMO claims that honesty of purpose and self-change are at the heart of its approach to water sector reforms, though some villagers thought otherwise. In Mokarshivangh village (WASMO-ERR) men reported how WASMO brought its own contractors to build their water supply works: "We told WASMO that their contractor will only be able to work from 8 a.m. to 5 p.m. so that we can monitor their work; also they could not bring any (raw) materials at night," (Khanjibhai, secretary of *pani samiti*, FGD, December 2007).

There is a common perception, rooted in arguments that essentialize women, that their participation in community institutions has a direct bearing on corruption, as they are known to work honestly: "If one-third of the members on the committee are women, it automatically brings the corruption down by 1/3rd, so it is essential to provide reservations to ensure that women are able to come up in public spheres," (interview with Hansaben, *pani samiti* member, Navagam village, Surendranagar district, September 2007). Another woman added: "Men try to figure out where they can make money through cuts in project allocations while women make sure that the work is done properly," (Prabhaben, a *pani samiti* member, Janada village, Bhavnagar district, April 2007).

Whither Water? Rethinking Decentralization, Facilitating Empowerment

According to the Network for Social Accountability (NSA) (2008) only 4.7 percent of rural habitations (as a percentage of 27,035 fully covered habitations) in Gujarat do not have access to drinking water. While this figure appears low, it is "high" when compared to states, such as Uttar Pradesh (0.4 percent not covered) or Karnataka (0.1 percent) which have had similar histories of sector reforms (e.g., the SWAJAL project in UP), albeit different trajectories. However, coverage figures "cover-up," literally, questions of who has access, functioning of schemes, population increase and water quality issues (WaterAid India 2005). One of the key points about Gujarat is that after the gradual phasing out of Dutch bilateral funds (by 2005), the various sector reform programs have been almost entirely funded by the state and central governments. Although WASMO claims that finances have never been a constraint to progress, by April 2008 only 58.78 percent of the total number of villages in the state (18,311) had *pani samitis* (see www.wasmo.org).

But the question is not one of numbers alone, impressive as they may sound. Instead as this chapter argues, while acknowledging the progress that has been made, there are still miles to go and gaps to bridge before women and marginalized communities can have equal access to sufficient and good quality water for drinking and personal hygiene as well as participate in community institutions and decision-making processes, including for women, within the household. Creating institutional space, while an important beginning, is not in itself enough as the "construction of interest and the construction of voice" requires engagement from below (Heller 2001), where politics is far more embedded in local traditional structures of power and planning capacity is weak (both in Panchayats and *pani samitis*). Civil society, NGOs and grassroots groups play an important role in facilitating a "new water commons" (Bakker 2007), but they work within constraints imposed by a techno-rationale framework from the top and are compelled to balance conflicting interests of caste, gender and increasingly faith in complex micro-communities. There are also inherent assumptions that stereotype women in specific roles, not only in terms of water management, but in the wider public domain, making gender-just change a challenge.

Decentralization is not a panacea for democratic environmental governance or sustainable water management. However, it presents us with a potentially transformative and empowering agenda, challenging us to reconstruct citizenship rights and define a new politics of water, equity and gender justice in a fractured society where the political articulation of human rights as livelihood rights (and vice versa) is critical. Rebuilding an inclusive, collective identity around water requires that we do not do away with the state, as the neoliberal language of sector reforms tends to suggest, but find effective strategies of making water governance work, particularly at the "messy meso level" (Cleaver and Franks 2005) where the institutional vacuum between policy and practice is most pronounced.

≈

Notes

1. The project was supported by Canada's International Development Research Centre (IDRC) as part of their global program on decentralization and gender rights.
2. The research was jointly undertaken in the states of western India, viz., Gujarat and Maharashtra, by a team based at Utthan, Ahmedabad, Gujarat, and by researchers at the Society for Promoting Participative Ecosystem Management (SOPPECOM), Pune and Tata Institute of Social Sciences (TISS), Mumbai, both in Maharashtra. I would like to deeply acknowledge the intellectual support of Seema Kulkarni, Project Leader, SOPPECOM, in helping develop ideas expressed in this chapter;

Yuthika Mathur and Dinesh Makwana for their support with the field research; our non-governmental organization (NGO) partners and women and men water users for their time.

3. Rondinelli et al. (1989) distinguish between *deconcentration* as the transfer of certain planning, financing and management tasks to local units of central agencies, without any inherent transfer of authority; *delegation* as the transfer of decision-making power to organizations that have semi-independent authorities, but are accountable to central bodies; and finally, *devolution* as the actual transfer of authority and power to local governments.
4. The Gender Equity Policy was the result of a long collaborative process between the state and civil society in Gujarat, supported by the United Nations Fund for Population Activities (UNFPA).
5. Sanitation has always been underfunded compared to drinking water schemes, and progress in meeting the sanitation targets under Millennium Development Goals (MDGs) is limited, partly because of the lack of political will (see WaterAid India 2005).
6. Insights in this section are based on extensive interviews with female *pani samiti* members and focus group discussions (FGDs) with both female and male members (separate and mixed), as well as non-members, often at water collection points along village transects in 13 villages from three institutional contexts: the WASMO-Ghogha project, WASMO-ERR project and NGO-facilitated decentralized water management.
7. Three villages from the Ghogha project area and five from the ERR project.
8. Five villages facilitated by local NGOs and the state-level water network, Pravah.
9. SHGs have been promoted by many NGOs to support savings and loan activities by poor women who find it difficult to access formal banks without adequate collateral (e.g., assets, such as land in their name). While the merits of SHGs have been widely debated, they were also meant to serve as entry points to facilitate women's mobilization and empowerment.
10. Without adequate benchmarks, it is difficult to assess this, except on the basis of perceptions of change.
11. In India, it is estimated that the national cost of women fetching water is 150 million woman work days per year, equivalent to a national loss of income of INR 10 billion or US$ 208 million (WASMO 2007: 1).

References

Agarwal, B. 2001. "Participatory Exclusions, Community Forestry and Gender: An Analysis for South Asia and a Conceptual Framework," *World Development*, 29(10): 1623–48.

Ahmed, S. 2002. "Mainstreaming Gender Equity in Water Management: Institutions, Policy and Practice in Gujarat, India," in S. Cummings, H. van Dam and M. Valk (eds), *Natural Resources Management and Gender: A Global Source Book*, pp. 33–43. Oxford: Oxfam Publishing.

Ahmed, S. (ed.) 2005. *Flowing Upstream: Empowering Women through Water Management Initiatives in India.* New Delhi: Foundation Books.

Asthana, A. N. 2004. "Corruption and Decentralisation: Evidence from India's Water Sector." Paper presented at the 30 WEDC International Conference, October, 25–29, Vientiane, Lao People's Democratic Republic.

Bakker, K. 2007. "The 'Commons' Versus the 'Commodity': Alter-globalisation, Anit-privatization and the Human Right to Water in the Global South," *Antipode*, 39(3): 430–55.

Barot, N. 1997. "A People's Movement towards Creating Sustainable Drinking Water Systems in Rural Gujarat," in N. Rao and L. Rurup (eds), *A Just Right: Women's Ownership of Natural Resources and Livelihood Security*, pp. 470–86. New Delhi: Friedrich Elbert Stifung.

Chhatre, A. 2007. "Accountability in Decentralisation and the Democratic Context: Theory and Evidence from India." Working Paper 23, World Resources Institute, Washington DC.

Cleaver, F. 1999. "Paradoxes of Participation: Questioning Participatory Approaches to Development," *Journal of International Development*, 11: 597–612.

Cleaver, F. and T. Franks. 2005. "Water Governance and Poverty: A Framework for Analysis, Bradford Centre for International Development." Research Paper 13, http://core.kmi.open.ac.uk/display/5664 (accessed September 15, 2013).

Coles, A. and T. Wallace (eds). 2005. *Gender, Water and Development.* Oxford: Berg.

Das, K. 2007. *Rapid Assessment of the Community Managed Water and Sanitation Programme in the Earthquake Affected Villages of Gujarat: A Report.* Ahmedabad: Gujarat Institute of Development Research.

Government of India (GoI). 2001. *Census of India*, http://www.censusindia.gov.in/2011-common/censusdataonline.html (accessed September 15, 2013).

Green, G., S. Joekes and M. Leach. 1998. "Questionable Links: Approaches to Gender in Environmental Research and Policy," in C. Jackson and R. Pearson (eds), *Feminist Visions of Development: Gender Analysis and Policy*, pp. 259–83. London: Routledge.

Grijpstra, B. 2006. "The Ghogha Project, 1997–2005: Perceptions and Reflections." Paper presented at the National Conference on Scaling Up Sector Reforms: Looking Ahead, Learning from the Past, April 28–29, Water and Sanitation Management Organisation (WASMO), Gandhinagar.

Gujarat State Drinking Water Infrastructure Corporation Limited (GSDWICL). 2000. *Gujarat Jal-Disha 2010: A Vision of a Healthy and Equitable Future with Drinking Water, Hygiene and Sanitation for All.* Ahmedabad: GSDWICL.

Hardiman, D. 1998. "Well Irrigation in Gujarat: Systems of Use, Hierarchies of Control," *Economic and Political Weekly*, 23(25): 1533–1544.

Harvey, D. 2003. *The New Imperialism.* New York: Oxford University Press.

Heller, P. 2001. "Moving the State: The Politics of Democratic Decentralisation in Kerala, South Africa and Porto Alegre," *Politics and Society*, 29(1): 131–63.

Iyer, R. R. 2007. *Towards Water Wisdom: Limits, Justice, Harmony.* New Delhi: Sage.

Kumar, M. D. and J. Talati. 2000. "Mitigating Drinking Water Crisis in Rural Gujarat: Seeking Technological and Institutional Options," *Water and Energy International*, 57(2): 43–54.

Mahatma Gandhi Labour Institute. 2004. *Gujarat Human Development Report 2004.* Ahmedabad: Mahatma Gandhi Labour Institute.

Mollinga, P. P. 2006. "IWRM in South Asia: A Concept Looking for a Constituency," in P. P. Mollinga, A. Dixit and K. Athukorala (eds), *Integrated Water Resources Management: Global Theory, Emerging Practice and Local Needs*, pp. 21–37. Water in South Asia, vol. 1. New Delhi: Sage.

Network for Social Accountability (NSA). 2008. "Public Provisioning for Rural Water Supply in India: Financial Needs Vs Government Commitments," http://www.nsa.org.in/Policybrief/314ruralwaterpage-3.htm (accessed September 15, 2008).

Oakley, P. et al. 1991. *Projects with People: The Practice of Participation in Rural Development.* Geneva: International Labour Organization (ILO).

O'Reilly, K. 2008. "Insider/Outsider Politics: Implementing Gendered Participation in Water Resource Management," in B. P. Resurreccion and R. Elmhirst (eds), *Gender and Natural Resources Management: Livelihoods, Mobility and Interventions*, pp. 195–212. London: Earthscan.

Prajapati, S. 2005. *Case Studies on Conflicts in Surendranagar Programme Area.* Ahmedabad: Aga Khan Rural Support Program (AKRSP) (I).

Ramachandraiah, C. 2001. "Drinking Water as a Fundamental Right," *Economic and Political Weekly*, 36(8): 619–21.

Reynders, J., S. Ahmed, J. Damodaran and J. Abbott. 2004. *Which Water? Ghogha Rural Water Supply and Sanitation: Final Project Progress Assessment.* New Delhi: Royal Netherlands Embassy.

Ribot, J. C. 2003. "Democratic Decentralization of Natural Resources: Institutional Choice and Discretionary Power Transfers in Sub-Saharan Africa," *Public Administration and Development*, 23: 53–65.

Rocheleau, D., B. Thomas-Slayter and E. Wangari (eds) 1996. *Feminist Political Ecology: Global Issues and Local Experiences.* London and New York: Routledge.

Rondinelli, D., J. McCullough and R. W. Johnson. 1989. "Analysing Decentralization Policies in Developing Countries: A Political-Economy Framework," *Development and Change*, 20(1): 57–87.

Rowland, J. 1997. "What is Empowerment?," in H. Afshar and F. Alikhan (eds), *Empowering Women for Development*, pp. 46–61. Hyderabad: Booklinks Corporation.

Sama, R. K. and I. Khurana, 2006. "WASMO: A Model for Implementation of the 73rd Constitutional Amendment." Paper prepared at the National Conference on Scaling Up Sector Reforms: Looking Ahead, Learning from the Past, April 28–29, WASMO, Gandhinagar.

Singh, N., P. Bhattacharya, G. Jacks and J. E. Gustafsson. 2004. "Women and Modern Domestic Water Supply Systems: Need for a Holistic Perspective," *Water Resources Management*, 18(3): 237–48.

United Nations Children's Fund (UNICEF) and World Wildlife Fund (WWF). 1997. *Fresh Water for India's Children and Nature*, New Delhi: UNICEF and WWF.

Water and Sanitation Management Organisation (WASMO). 2008. *Five Years of Decentralised Community Managed Water Supply Programme*. Gandhinagar: WASMO.

———. 2007. *Empowering Women for Improved Access to Safe Water*. Gandhinagar: WASMO.

WaterAid India. 2005. *Drinking Water and Sanitation Status in India: Coverage, Financing and Emerging Concerns*. New Delhi: WaterAid.

Women's Environment & Development Organization (WEDO). 2003. *Untapped Connections: Gender, Water and Poverty: Key Issues, Government Commitments and Actions for Sustainable Development*. New York: WEDO.

3

Rethinking Gender Inclusion and Equity in Irrigation Policy

Insights from Nepal

Pranita Bhushan Udas

Gender equity is an important component of the social equity agenda of development programs and plans. Gender equity, as defined by the United Nations is: "fairness of treatment for women and men, according to their respective needs. This may include equal treatment or treatment that is different but considered equivalent in terms of rights, benefits, obligations and opportunities" (UNESCO 2000). Improving equity is viewed as a central concern of the state. Most national constitutions promise equal basic rights and freedom for all citizens and most explicitly promise to safeguard equality regardless of age, caste, race or sex (King 2001). The state as the supreme governing body imposes rules of equality on its citizens as a way to eliminate discrimination. For instance, the Millennium Development Goals (MDGs) adopted by many governments including Nepal are significant as they are a composite of development goals of the international community. The MDGs point to the need for recognition of gender as a cross-sectoral dimension because all MDGs are intrinsically linked to gender equity (Khosla 2003: 14). However, the term "fairness" in the definition of gender equity is vague. What is considered "fair" is culture-specific and context-specific. In addition, making a policy commitment for gendered change is a novel effort. However, bringing changes in gender relation is equally complicated.

This chapter analyzes policy interventions to address gender equity in the irrigation policy of Nepal. A review of irrigation water policy documents and the evidence gathered from fieldwork on its implementation show that the realization of gender equity is a complex process. Realization of gender equity as a goal in the drafting of irrigation policies and their implementation has been limited to visibilizing women members in water user' committees. The policy documents (water resources acts, irrigation policies and regulations) have failed to provide guidelines on "fair treatment" as mentioned in the UN definition. This chapter is an attempt to explore and explain the reasons for these limitations in policy-making and policy-implementation in the context of Nepal. The main argument in this chapter is that there are gaps in the creation of context-specific knowledge for providing working definitions of gender equity in specific areas of water management such as irrigation. The justification for including the gender dimension in the irrigation sector is typically limited to arguments for gender equality in development, i.e., arguments based on the idea of equal population of males and females in a society and, therefore, the need for equal representation. Beyond making this argument, there has not been any attempt on the part of policy-makers to address issues, such as making labor contribution towards construction of irrigation canals equitable and proportionate to landholding size, or limiting membership criteria to land ownership, etc. This chapter argues that addressing such issues that help female and male water users (from different socio-economic backgrounds) to access water resources in a fair manner can bring about gender equity in water sector.

The chapter is based on my doctoral research titled "Gendered Participation in Water Management in Nepal: Discourses, Policies and Practices." The field study was carried out between 2004 and 2007 at two levels; one, study of irrigation systems located in Baruwa river basin in the Udayapur district of eastern Nepal and second, study of implementers at the sub-divisional level, in the regional and central offices of the department of irrigation. The data collection method included participatory observations, focus group discussions (FGDs), interactive workshops with officers and interviews.

The introductory section of the chapter is followed by three sections. The first section briefly reviews the state of progress in institutionalizing gender equity issues in water sector. The second section details the empirical findings and is divided into four sub-sections. The first sub-section provides an overview of gender relations in Nepali society and integration of gender in the development plans of Nepal in general; the second sub-section reviews the way the issue of gender equity is addressed in the current irrigation policy; the third sub-section presents irrigation officers' perspective on implementing gender equity in irrigation policies and the associated complexities and challenges; and the fourth sub-section addresses gender equity issues at the farm level. The third section is a concluding section that identifies gaps in policy prescriptions and practices.

Conceptualizing Gender Equity in Irrigation Sector

Access to irrigated water is important for enhancing agricultural productivity. Increased productivity, in turn, can lead to well-being and enhanced economic growth. The access to irrigation water is conditioned by three factors. First, in a particular location it is defined by water usage rights. Often, water usage rights in an irrigation system are attached to the investment made in the construction of irrigation infrastructure and thus are accessible to those who can invest, particularly landlords and rich peasants. Economically poor sections of the society are mostly deprived of access to water for irrigation. Second, the public investment in irrigation sector often imposes inequity, instead of bridging the gaps between "haves" and "have nots." The return of public investment in irrigation sector benefits the population within the command area according to the landholding size, with large landholders having a bigger share of benefits than small landholders and the landless being completely excluded. Third, irrigation sector is dominated by male engineers and irrigation activities at the farm level are understood as works requiring physical labor of males, such as construction of canals and head-works. This excludes and makes women invisible in the water sector (Zwarteveen 2006; Liebrand 2010).

Management of irrigation water comprises of (*a*) water-related activities and (*b*) property rights and associated duties (Martin and Yoder 1987). Water-related activities include acquisition, allocation, distribution and drainage, as also activities related to "structures" for controlling "water," like design, construction, operation and maintenance. Further, they include activities related to "organization" of people who manage water and structures, like decision-making, resource mobilization, communication and conflict management (Uphoff et al. 1985). Property rights related to irrigation structures, land and water are important as they define access to these and the associated obligations. To be precise, Beccar et al. (2002) define irrigation system as a complex set of *physical elements*, i.e., water sources and flows, the places where it (irrigation system) is applied and the hydraulic infrastructure to catch, conduct and distribute it (water); *normative elements*, i.e., rules, rights and obligations related to access to water and other necessary resources; *organizational elements*, i.e., human organization to govern, operate and sustain the systems; and *agro-productive elements*, i.e., soil, crops, technology, capital, labor force, the capacities for and the knowledge of the art of irrigation. The organization and management of irrigation-related activites are shaped by the pre-existing power relations in agrarian societies.

In the context of irrigation water management, water rights are defined as authorized demands to use (part of) a flow of water. There are certain privileges, restrictions, obligations and sanctions accompanying this authorization, among which a key element is the power to take part in collective decision-making about the management of irrigation system. The main element of this definition is authorization, i.e., rights can be exercised only when water use is certified by an authority (individual or collective) with legitimacy and power of enforcement, recognized by users and non-users alike (Beccar, et al. 2002).

Gender perspective in irrigation sector emerged with tail-enders'[1] right to have access to water and their plight on not receiving water in time. Understanding their needs and interests, and power relations in the system can be considered to be an entry point that raises debate on the gender dimension in water sector (Zwarteveen 1995). The water distribution in canal irrigation system is based on gravitational flow. The upstream users thus receive water first and tail-enders at the end (Ghimire 2005). Women-headed households among the tail-enders are the worst affected by unequal water distribution in the system.

The need to address gender imbalance in existing irrigation programs is justified by such factors as the significant role women play in farm activities, the different needs of women and women for irrigation water (Zwarteveen 1995; van Koppen et al. 2001) and the need for increasing project efficiency since, for instance, some irrigation projects failed due to gender biases in irrigation project design (Bruins and Heijmans 1993). In other cases, irrigation projects are found to have differential impact on male and female farmers. Mehta (1996) has shown increase in women's workload in rural India after the irrigation project. Van Koppen (2002) concludes that gender is an important issue to be considered in irrigation sector. However, the challenge is to answer "how"?

Gender bias refers both to unequal access to resources (land, water, credit, etc.) and to gender-differentiated access to the process of making and implementing decisions. What is important is not "who does what," but the exclusiveness of role distribution and its implications for resource allocation and distribution of power (Zwarteveen and Bennett 2005: 14) that leads to the well-being of one at the cost of the other. It is these biases that the discourse on gender equity aims to reduce. It refers to the fact that women and men play different roles and have different needs and interests. The possible measures that allow equitable access and control over water resources can help bridge the gap between "have" and "have nots."

Insights from Irrigation Sector in Nepal

This section brings forth empirical evidence to substantiate and understand gender equity, irrigation practice and policy linkages.

Gender Relations in Nepalese Society and Gender Development Debate

Gender relations across communities in Nepal cannot be generalized. Gender norms and values of the Nepalese vary according to the geographical location, caste and class dynamics. For example, the difference in gender-specific roles and responsibilities among hill inhabitants in the north is relatively less than that among those communities living in the southern plains of Terai.[2] The spatial divide between men and women in Terai (private as women's domain and public as men's) is more than that among hill inhabitants. A group called *matwali*, a liquor-drinking Tibeto-Burmese community (speaking a language belonging to a sub-family of Sino-Tibetan languages that includes Tibetan and Burmese), gives equal importance to sons and daughters in family rituals.[3] On the other hand, the "Aryans" that include the Brahmin and Chetri castes have different roles for sons and daughters in rituals related to birth, marriage and death. These gender differences tend to define gender relations in the public sphere to a certain extent. Thus, irrigation-related activities cannot be considered in isolation from these prescribed gender roles. What matters the most is where the activity has taken place, with whom and how it will be done. These determine who can or cannot be involved in irrigation activities, such as canal cleaning, irrigating farms or attending meetings.

The gender inequality in access to irrigation water in Nepali society became prominent with the introduction of hereditary *varna* (caste) system in 1361 by the then king Jayasthiti Malla. This system classified people into different caste groups (Singh 2006: 122) and two major clusters emerged on the basis of norms about water use: "touchables" and "untouchables." "Touchables" were high-caste people from whose hands drinking water was permissible, and "untouchables" were low-caste people from whose hands drinking water was not permissible (Amatya 2006: 4).

The inequities and inequalities in society reached their peak under the influence of *Kulin* system (practice of polygamy among certain kinds of Brahmins considered *Kulin*s and consequent degradation of the women wedded to these polygynous *Kulin* husbands) of Bengal and *Sati* (the custom of burning a widow on the funeral pyre along with her husband's dead body) brought in from India. The earlier relatively equal gender relationships thus turned unequal with men being considered superior and women their subordinates (cf. Singh 2006: 40). Women were and are, even today, considered untouchables during certain periods of biological changes, such as menstruation and the first few days after childbirth. In addition, widows and infertile women receive ill treatment throughout their lives.

To address the issue of gender imbalance in the society, Nepal government introduced policies and plans for women especially after its participation in the first International Women's Conference in Mexico in 1975. The United Nations Declaration on Women's Rights of 1975 encouraged the governments to craft national policies addressing gender discrimination. As a result, in 1977, the Women Services Coordination Council under the Social Services Coordination Council was established in Nepal. In addition, Nepal signed a number of international instruments like Convention on Elimination of All Forms of Discrimination Against Women (CEDAW), International Covenant on Civil and Political Rights (ICCPR), International Covenant on Economic, Social and Cultural Rights (ICESCR), Convention Against Torture (CAT), Convention on the Rights of the Child (CRC) and Convention on Elimination of All Forms of Racial Discrimination (CERD). Factoring in its commitment to the Beijing Declaration of 1995, the Nepal government also formulated a Gender Equity and Women's Empowerment National Work Plan in 1997 encompassing 12 sectors. They are women and poverty, education, health, violence, armed insurgency, economy, policymaking, institutional structure, human rights, environment and children (HMG/Nepal 1998).

Nepal's national plans and policies also addressed gender concerns over time. The Sixth Five Year Plan (1975–80), for instance, recognized women for the first time as equal shareholders of development and focused on increasing women's skills and involving women in income-generating activities. Consequently, the subsequent plans included "plans for women" as a separate chapter. The Seventh Plan (1985–90) emphasized the introduction of a quota for women in trainings and other development activities. The Eighth Plan (1992–97) focussed on policy amendments to ensure women's participation. As a result, irrigation policy of 1992 guided increasing women participation in Water Users' Association (WUA) committees. The Ninth Five Year Plan (1997–2002) emphasized the integration of women into the mainstream of development, particularly institutional development and empowerment. The Tenth Plan (2002–07), which is also a poverty reduction strategy paper (PRSP) of the country, introduced quantitative targets to monitor outcomes of programs aimed at the upliftment of women. It also emphasized on gender-sensitive budgeting, evaluation and monitoring.

Gender Concerns in Irrigation Policy

The irrigation sector in Nepal is dominated by funding agencies, as more than half of its budget is met from loan and grants (Shah and Singh 2001). In this context, water policy of Nepal has addressed the issues of social equity to some extent. For example, Nepal is a country where the definition of large, small and medium irrigation system varies with the geography. The system that is small for Terai is large for hill communities. Similarly, the labor contribution of farmers from different regions and government support for canal rehabilitation varies with location. The remoteness, poor economy and hardships of the areas that people live in, are considered while deciding farmers' contributions towards the rehabilitation or construction of any irrigation system with government support (HMG/Nepal 2003a and b, 1992).

The component of gender equity was incorporated in the irrigation policy of 1992 following the guidelines of the Eighth Plan. Accordingly, the irrigation policy of 1992 called for minimum 20 percent participation of women users in the WUA committees to ensure their minimum participation in formal decision-making forums. This provision was revised in the irrigation policy and regulation of 2003 and the participation was increased to 33 percent. In addition, the women development office located in each district was made responsible for coordinating and mainstreaming gender issues across sectors in 2005. Each government office in a district was to select one of its staff members as gender focal person. The assumption is that the focal person would be trained to mainstream gender issues in the activities implemented by the government office.

Irrigation Missionary: Translating the Policy Objectives

In the department of irrigation, 84 percent are civil engineers as core staff, another 15 percent are engineers from non-civil backgrounds and the remaining one percent are sociologists and association organizers, i.e., professionals with education in social sciences (Timilsina et al. 2008). Sociologists and association organizers are the two categories of staff in the organizational structure of the department responsible for the social aspect of irrigation, often termed as the 'software' of the irrigation projects. However, the dominance of civil engineers in the department makes it appear as an organization for the construction of irrigation systems. At the local or field level, even after 20 years of policy efforts to increase women's participation in the WUA committees, the fact is that women's visibility in the users' organizations is nominal. An analysis of 588 WUAs in 31 districts registered with the National Federation of Irrigation Water Users' Association in 2006 showed that only four districts had more than 20 percent women members on an average in the committees.

This section attempts to understand officers' perspective on addressing gender concerns in the country's irrigation policy. Table 3.1 presents an overview of gender concerns in the irrigation policy over time. The information in this section is based on interactive discussions, interviews and field visits with officers at district offices in Udayapur, regional offices in Biratnagar, western Nepal, and at the central department of irrigation in Kathmandu. Both technical and non-technical officers were consulted for this study. Officers shared their experiences of implementing national policy and programs in project modes. The programs implemented in project mode are guided by the rules of funding agencies under the guidelines of national irrigation policy. The implementation modalities vary slightly from one project to the other.

Table 3.1
Policy clauses on representation of women and excluded sections

Irrigation Policy, 1992	It mentions that necessary emphasis shall be given to the provision that there shall be at least 20 percent female users in all the executive units of WUAs.
Irrigation Policy, 1996 (Amended)	It mentions that necessary efforts have to be made to make sure 20 percent of the total executive members of the WUAs are women.
Irrigation Regulation, 2000	It legalized the policy objective on women's participation. It mentions, "The users desirous to use any irrigation system developed and operated by His Majesty's Government shall be required to constitute a users' association having the executive committee of not exceeding nine members including at least two women members."
Irrigation Policy (Amended), 2003	It provides guidelines for drawing 33 percent of the members of WUAs from women and giving equal representation to backward caste groups and minorities.
Irrigation Regulation (Amended), 2003	Thirty-three percent of members of the WUAs shall be women, and there shall be representation of *dalit*, downtrodden and backward ethnic communities in such associations.

Source: Review of respective policy documents by the author.

The interaction organised as an interactive meeting with officers on November 31, 2007 to share their experience on implementing gendered immigration policy clearly revealed the concerns about the ways in which gender gap is addressed in irrigation programs. Officers concluded that increasing women's participation in the users' committees is a provision mentioned in the national irrigation policy to address gender issues. Therefore, the discussion and debate focused on achieving increased women's participation. However, few respondents expressed doubts on whether only women can represent women's concerns. Case 1 presented below shows concerns and learnings on implementing the gender component of the irrigation policy.

Case 1

A sociologist during the interview held on November 31, 2007, explained a case of second irrigation sector project (SISP),[4] which had the provision for recruiting female field-level staff as community organizers. As a result, three female community organizers were selected in his working area. However, the evaluation of the performance of female community organizers after a year indicated poor results. One of the community organizers got married and left the village while the other two could not continue due to their household and family commitments. This made him rethink the utility of recruiting female mobilizers to address gender concerns better. In his experience, these women despite residing in the village that came under his working area could not play their role effectively as they had to handle both household responsibilities and their new assignment as community mobilizers. This shows that no effort was made to change the gender disparities these women faced in their day-to-day activities. Therefore, he concluded:

> Irrigation regulation that has provisioned 33 percent women's representation in the committee and SISP policy that recruited women mobilisers are important to increase women's participation in formal irrigation world. However, to achieve the objective, there is a need for other programs on empowerment and altering existing gender roles. These programs can be implemented either by irrigation department or any other line agencies, but has to go hand in hand with irrigation project construction.

Case 2

An engineer during the interview held on November 31, 2007, opined that women users participated in water projects only when their primary needs were addressed. To support the statement he shared his experience of implementing Dhaulagiri irrigation project financed by International Labour Organization (ILO). At first, the project included women empowerment components, such as literacy classes and women's saving group, in addition to construction of irrigation systems. Despite this, women's participation and visibility in the irrigation users group was nominal. After an evaluation, the second phase included community development and environmental protection components whereby Community Development and Environmental Protection Committee, as well as income-generation groups of women, were formed. Along with this, biogas plants for cooking were also installed. These components had a positive impact on encouraging women to come forward and participate in the implementation of irrigation programs, because these addressed women members' primary needs (cooking and income generation).

Case 3

The chief of one of the Irrigation Management Transfer Project (IMTP) in large irrigation projects of Terai during the interview held on November 31, 2007, argued that economic incentives is an important factor for women to break the traditional norms that keep them away from formal water forums. Kamala irrigation project, one of the large irrigation systems in Nepal's Terai region, aimed to increase women's participation as part of the IMTP.[5] It adopted affirmative action like recruitment of a female sociologist, female farmer organizers and local consultants, which yielded positive outcomes to some extent. Since the post of female farmer organizer was a paid one, women were encouraged by family members to apply for it. The rule was to recruit unmarried women, but it was found out that even married women applied for the post pretending that they were not married because of the lucrative remuneration involved. In addition, the rate of service fee collection increased considerably with the involvement of women organizers. He argued that the economic incentive is the main driver that encouraged women and their family members to participate in the activities of WUAs although most of these women were from families where they had to be under the veil.

Gender Inequity Issues in Water Use Practices at Farm Level

Evidence from farm-level water management showed that gender inequity at the farm level is linked to issues of social inequity. Addressing gender inequity at the farm level thus implies working on the issues of social inequity. The afore-discussed cases of three irrigation systems studied for a year explain the linkages by tracing women's participation in WUAs and their contribution (in terms of labor) towards the irrigation systems. The participatory irrigation policy calls for a certain percentage of users' contribution while receiving government support for system rehabilitation. The irrigation policy of 1992 factored the remoteness, poor economy and the hardships of life in the mountainous and hilly areas into the issue of farmers' contribution towards the rehabilitation or construction of new irrigation systems with government support (HMG/Nepal 1992). However, it did not address the problems faced by poor farmers in a certain geographical location. Therefore, the 2003 amendments in the irrigation policy and regulation attempted to address these issues by making the water users' contribution proportionate to their land holding size irrespective of the geographical location. However, in practice, the labor contribution of farmers was found to be non-proportionate to the size of irrigated lands they held. It is because traditionally all the farmers used to collectively clean the canal before rainy season to bring water into the canal. Hence, the labor contribution was irrespective of the size of their respective irrigated landholdings. After the rehabilitation of irrigation system by the government, the total labor required for maintaining the system has lessened. The irrigation system in Arghali and Cherlung in western Nepal are considered to be examples of success in bringing about social equity in water management which was characterized by resource mobilization, whereby each of the member households sent one laborer, irrespective of the amount of water allocated to them solely for emergency maintenance purposes (Martin 1986: 301). Customarily, the labor required for an irrigation system was intensive and required only once and that too for a specific time period. Thus, the quantum of labor contribution to the cleaning of the irrigation canal was often equal for all the users. The irrigation policy of 2003, while calling for contribution proportionate to the size of irrigated land of each contributing household, does not explicitly explain whether the contribution is in terms of a cash payment or labor. Thus, in practice, water users in a command area contribute an equal amount of labor irrespective of the varying sizes of the irrigated land held by each, whereas the cash contribution is found to be proportionate to the size of irrigated land. This has implication for small landholders, especially women, who receive fewer benefits from the irrigation system than do large landholders.

Three irrigation systems, viz., Upper Baruwa, Bhusune Asare and Baruwa, receive water from Baruwa River in Udayapur[6] district and were rehabilitated with government support at different points of time. The first and the third systems were initiated by farmers. As per the participatory policy, at the time of system rehabilitation, the irrigators had to make contributions amounting to 15 percent of the total budget for rehabilitation in order to receive the government grant. To begin the rehabilitation work and express water users' commitment towards system rehabilitation, the WUA had to deposit 5 percent of the total amount in bank as upfront contribution. The remaining 10 percent of the contribution to be made could be in the form of labor, any other kind of service or cash.

In all the three systems, labor contribution was made on the basis of the number of households, irrespective of the size of irrigated lands they held, though the cash contribution as initial deposit was based on the amount of land irrigated. During the field study, I observed the rehabilitation process of Bhusune Asari irrigation system. The meeting took place in the community hall constructed by forest users' group of the community, among whom many of them were members of this irrigation system. The user who came first had to wait for an hour to start the meeting, as all users did not arrive on time. There were about 40 users, out of which nine were women. One of the female users raised the issue of labor contribution proportionate to irrigated land. Few men also supported her. However, their voices appeared minor and evaporated as the committee members and other representative farmers did not pay attention. Similar observations were made in irrigation systems studied in 2001 in central Nepal (Udas 2002).

Further inquiry from the woman who raised the concern revealed her view that they as small landholders were losers if all the users had to contribute equal amount of labor, irrespective of the benefits one got from the system. She expressed concern that irrigation system brought more wealth to large landholders and so they should contribute more in terms of labor. It was further revealed that women from poor households attended the WUA meetings and contributed to canal cleaning and other activities of the WUAs because their male members would go out of the villages for work to earn higher wages. The low wage rate for women members in local labor market was the reason why women from poor households contributed to the labor-intensive activities of the WUAs. WUA meetings were time-consuming and not economically beneficial to the households with small irrigated plots. Hence, such households often decided that their womenfolk should represent the household for labour work and the male members should go to work in market, because it fetches him more money. Thus, it was not in the interest of women members or the policy incentives that encourage women's participation in WUAs, but rather is a household's decision to maximize their opportunity cost of participating in the WUA meetings.

Two issues related to women's participation and their labor contribution are raised here. The women who participated in the water users meetings did not do so because of policy provisions that were meant to increase women's participation. They did so because of their families' decision to maximize their benefits. Further, the irrigation policy remains silent on the issue of labor contribution and does not squarely address the issue of social inequity, thereby perpetuating the perception of women's labor as unpaid work.

Policy Prescription and Practice: Exploring the Missing Link

Irrigation regulation and act is a binding document though its effectiveness depends on the way it is formulated and implemented. Inequities, on other hand, are linked to power relations, bargaining, subordination, domination and exclusion. Working towards removing inequities at the community level requires social mobilization and action, which might take a long time if is expected to be initiated by community members alone. State policies framed to address inequities can facilitate the process of social action required to bridge the gaps in gender equity.

The study of irrigation policy and its implementation in Nepal shows that the prescription at the moment is limited to increasing women's visibility in order to address the concern about gender inequity. Officers implementing the policy and working towards achieving minimum participation of women users in the WUAs for more than 20 years conclude that the present policy provision is not enough to address gender inequities. They suggest that to achieve increased women's participation, there is a need for associated activities and incentives that factor in women's water needs. The observations from fieldwork indicate that women's participation, instead of being the result of policy provisions, is conditioned by their households' decision which, in turn, is based on their evaluation of the opportunity costs of participation. On other hand, the persisting gender inequities in accessing water and receiving equitable benefits from irrigation systems are linked to inequities in labor contribution.

That labor contribution towards maintaining irrigation systems was irrespective of the size of irrigated lands held by the contributing households has been observed not only in the systems discussed in this essay. In an earlier study of farmer-managed irrigation systems, E. D. Martin mentions unequal labor contribution with respect to water allocation as an instance of systemic inequity (1986: 301). Though the issue of labor contribution does not appear to be a gender issue on the surface, the fieldwork shows that it has links to gender inequity. It also shows that gender issues are an integral part of the social equity issues. Thus, the policy prescription that labor contribution be proportionate to the size of irrigated land is one of the important ways to minimize gender inequity in the irrigation sector of Nepal.

Margreet Zwarteveen (2000) argues that in order to have a public forum (like WUAs) in which water users and irrigators can deliberate as peers, it is necessary to eliminate systemic social and gender inequalities. The evidence for unequal labor contribution found in this study is one example of the many water-related activities that perpetuate social inequities in general and gender inequity in particular. Hence, mapping of such inequities in the irrigation sector could help in working out a practical definition of gender inequity. The present policy is aimed at ensuring women's participation in decision-making forums through quota system – an intent which is extremely significant – but to realise it, exploring the roots of inequities in the sector and addressing them through policies are a priority. The strength of the quota system is that it brings changes in the dominant discourse on irrigation water management. For example, challenging the dominant notions, such as "public domain is only male domain" is found to be altered by policy effort to increase women's participation in committees (Udas 2006: 18). However, such policy provisions for women could be an entry point to provide space for those who are excluded, but is not an end in itself to achieve gender justice. With the example of unequal labor contribution, this essay aims to open a discussion on mapping other inequities in water-related activities and exploring their linkages to gender injustice. To conclude, unless gender in water is part of a broader social equity agenda, addressing gender concerns in a capsule will remain nominal and instrumentalist.

♒

Notes

1. Tail-ender denotes farmers who are located at the end of the command area in canal irrigation system and who receive water at the end.
2. Nepal is geographically divided into three zones. The low-lying southern plains next to the Indian border called the Terai with an altitude varying between 60 and 500 m above the sea level; the hills of Nepal with an

altitude varying between 500 to 2500 m above the sea level; and the mountains at 2500 m or more above the sea level (Gyawali 2001: 179).
3. People in Nepal can be broadly divided in two categories. The first are people of Indo-Aryan origins, who are known as non-*matwali* and who consider drinking liquor a crime. The second are of Mongoloid or Tibeto-Burmese origins, who are known as *matwali*, for whom liquor has high value and is important in rituals related to birth and death (Bennett 1983).
4. SISP (1996–2003) is a project implemented by the department of irrigation that aimed to rehabilitate farmer-managed irrigation system and construct new ones in the Eastern and Central Development region of the country. The project was funded through loans by the Asian Development Bank (ADB).
5. IMTP is launched in 1992–94 and 1995–2001. The project included systematic turnover of 21 irrigation systems mostly in Terai and few in hills. Under financial support from ADB, Nepal's Department of Irrigation (DOI) implemented the project.
6. Udayapur district spreads over Mahabharat range in north to Siwalik range in south within an area of 2063 sq. km and is located in inner Terai.

References

Amatya, S. 2006. *Water and Culture.* Kathmandu: Jalsrot Vikash Sanstha.
Beccar, L., R. Boelens and P. Hoogendam. 2002. "Water Rights and Collective Action in Community Irrigation," in R. Boelens and P. Hoogendam (eds), *Water Rights and Empowerment*, pp. 1–22. Assen, The Netherlands: Koninklikjke Van Gorcum BV.
Bennett, L. 1983. *Dangerous Wives and Sacred Sisters: Social and Symbolic Roles of High-caste Women in Nepal.* New York: Columbia University Press.
Bruins, B. and A. Heijmans. 1993. *Gender Biases in Irrigation Projects: Gender Considerations in the Rehabilitation of Bauraha Irrigation System in the District of Dang, Nepal.* Kathmandu: SNV.
Ghimire, Sujan. 2005. "Women and Irrigation in Nepal: Context, Issues and prospects." Occasional paper on Sociology and Anthropology 9, Central Department of Sociology Anthropology, Tribhuwan University, Kirtipur, Nepal.
Gyawali, D. 2001. *Water in Nepal.* Kathmandu: Himal Books.
HMG/Nepal. 2003a. *Irrigation Policy 2003.* Kathmandu: Department of Irrigation, Ministry of Water Resources.
———. 2003b. *Irrigation Regulation 2003.* Kathmandu: Department of Irrigation, Ministry of Water Resources.
———. 1998. *The Ninth Plan.* Kathmandu: National Planning Commission.
———. 1992. *Irrigation Policy 1992.* Kathmandu: Department of Irrigation, Ministry of Water Resources.

Khosla, Prabha. 2003. "Tapping into Sustainability: Issues and Trends in Gender Mainstreaming in Water and Sanitation." Background Paper presented at the "Gender and Water" session of the Third World Water Forum, International Water and Sanitation Centre, International Water and Sanitation Centre, IRC, the Netherlands.

King, E. M. 2001. *Engendering Development: Through Gender Equality in Rights, Resources and Voice.* World Bank Policy Research Report 21776. Washington DC: The World Bank and Oxford University Press.

Liebrand, J. 2010. "Masculinities: A Scale Challenge in Irrigation Governance in Nepal." Paper prsented at the Fifth Seminar of the Farmer Managed Irrigation Systems Promotion Trust, "Dynamics of Farmer Managed Irrigation Systems: Socioinstitutional, Economic and Technical context," March 25–26, Farmer Managed Irrigation Systems Promotion Trust, Kathmandu, Nepal.

Martin, E. and R. Yoder. 1987. "Institutions for Irrigation Management in Farmer Managed Systems: Examples from the Hills of Nepal." International Irrigation Management Institute (IIMI) Research Paper 5, IIMI, Colombo, Sri Lanka.

Martin, E. D. 1986. "Resource Mobilization, Water Allocation, and Farmer Organization in Hill Irrigation Systems in Nepal." PhD thesis, Cornell University, Ithaca, NY.

Mehta, M. 1996. "Our Lives Are No Different from that of Our Buffaloes," in Barbara Thomas Slayter and Easter Wangari (eds), *Feminist Political Ecology*, pp. 180–210. London and New York: Routledge.

Shah, Shree Govinda and Gautam Narayan Singh. 2001. *Irrigation Development in Nepal Investment, Efficiency and Institution, Policy Analysis in Agriculture and Related Resource Management.* Research Report Series 47, Winrock International, Kathmandu, Nepal.

Singh, Subodh Kumar. 2006. *The Great Sons of Tharus: Sakyamuni Buddha and Asoka the Great.* New Nepal Press, Kathmandu, Nepal.

Timilsina, U., S. Sijapati, J. Dahal, B. R. Adhikari, S. R. Karki, G. L. Upadhaya and A. Poudel. 2008. *Study Report on Human Resource Assessment of DOI.* Kathmandu: Water Managment Branch, Department of Irrigation, Ministry of Water Resources, Government of Nepal.

Udas, P. B. 2006. *Quota Systems and Women's Participation: Lessons from Water Policy in Nepal.* New voices, New perspectives Series, vol. 3 of United Nations International Research and Training Institute for the Advancement of Women, Santa Domingo. Available online at http://un-instraw.org/en/images/stories/NewVoices/nv-bhushan.pdf. (accessed February 2, 2007).

———. 2002. "Gender, Policy and Water Users' Association: A Case Study of WUA in Nepal." MSc. thesis, Wageningen University, Wageningen, The Netherlands.

UNESCO. 2000. Gender Equality and Equity. A summary review of UNESCO's accomplishments since the Fourth World Conference on

Women (Beijing 1995). Unit for Promotion of the Status of Women and Gender Equality, UNESCO, Paris. Available online at http://unesdoc.unesco.org/images/0012/001211/121145e.pdf (accessed October 15, 2013).

Uphoff, N., R. Meinzen-Dick and N. St. Julien. 1985. *Improving Policies and Programs for Farmer Organization and Participation in Irrigation Water Management.* Report prepared for the Water Management Synthesis Project, Cornell University, Ithaca, NY.

van Koppen, B. 2002. *A Gender Performance Indicator for Irrigation: Concepts, Tools, and Applications.* Research Report 59. Colombo, Sri Lanka: International Water Management Institute (IWMI).

van Koppen, B., J. v. Etten, P. Bajracharya and A. Tuladhar. 2001. "Women Irrigators and Leaders in the West Gandak Scheme, Nepal." IWMI Working Papers H028139, IWMI, Colombo, Sri Lanka.

Zwarteveen, Margreet and Vivienne Bennett. 2005. 'The Connection between Gender and Water Management," in Vivienne Bennett, S. Davila-Poblete and Maia Nieves Rico (eds), *Opposing Currents: The Politics of Water and Gender in Latin America*, pp. 13–29. Pittsburgh, PA: University of Pittsburgh Press.

Zwarteveen, Margreet. 2006. "Wedlock or Deadlock? Feminists' Attemps to engage Irrigation Engineers." PhD thesis, Wageningen University, Wageningen, The Netherlands.

———. 2000. "Access, Participatory Parity and Democracy," in D. Virchow and J. V. Braun (eds), *Villages in the Future: Crops, Jobs and Livelihood*, pp. 215–21. Berlin and Heidelberg: Springer-Verlag.

———. 1995. "Linking Women to the Main Canal: Gender and Irrigation Management." Gatekeeper Series 54, Discussion Paper. London: International Institute for Environment and Development (IIED).

4

Entering Male Domain and Challenging Stereotypes

A Case Study on Gender and Irrigation in Sindh, Pakistan

Shaheen Ashraf Shah and *Nazeer Ahmed Memon*

Introduction

The Millennium Development Goals (MDGs) include reduction of the proportion of people without sustainable access to safe water by half, promotion of gender equality and empowerment of women at all levels. The Dublin Statement on Water and Sustainable Development highlights women's central role in provisioning, management and safeguarding of water. There are many other good reasons for recognizing women's role in water management. Gender affects all aspects of socio-economic behavior and beliefs. Failure to take gender differences and inequalities into account can result in futile development projects in general and water management projects in particular (see GWA and UNDP 2006; Walker 2006; World Bank 2003; DFID 2000; Meinzen-Dick and Zwarteveen 1998).

Gender equality in water management is also consistent with the aims and principles of the Government of Pakistan, which recognizes women's social and economic rights in the National Plan of Action (NPA) and National Policy for Development and Empowerment of Women (NPDEW). But water sector in Pakistan is still often considered to be outside the purview of women. As such a limited number of women have become prominent in this area as planners, managers, technicians, researchers and professionals. They are still not adequately recognized as a party in the current national debate on dams, water distribution and competing demands for water within sectors. Further, most of the water projects rarely recognize the different water needs of women and the impact of women's participation on the efficiency of water-related projects.

This chapter highlights the fact that despite water reforms in the Sindh province of Pakistan, policies and institutions have been gender-blind, leading to a curtailment of women's participation. The chapter also argues that lack of women's participation has adversely affected the efficiency and effectiveness of water management. It concludes that current policies of devolving water management need to be increasingly responsive to women's specific water needs and interests, if they are to address efficiency as well as equity concerns. This study draws upon information from both secondary and primary sources. It is especially based on the information provided by organizations involved in the implementation of the water management reform program and 10 focus group discussions (FGDs) with women farmers and water users, randomly selected from head, middle and tail distributaries/minors of Nara Canal, where the reform program is being implemented. The women farmers belonging to different villages and affected by the reform program, range between 20 and 50 years of age, are mainly uneducated and work as agricultural laborers. However, two FGDs were conducted in detail with women water professionals working for the reform program, to find out more about the status of women's participation at institutional levels. The FGDs with these women revealed that they were well educated, both married and unmarried, possessed professional experience of 3–10 years and were working as social organizers for the reform program.

This chapter is divided into four parts. The first part focuses on the factors that determine women's participation in water management and the gendered decentralization of water resources in the local context of Sindh. The second part evaluates women's participation in the ongoing reform program at policy, institutional and community levels. The third part provides evidence from the field for how lack of women's participation and consideration of their water needs has impacted on the effectiveness of water-related infrastructures. The fourth part offers certain conclusions based on the evidence provided.

Why Women in Water Management in Sindh?

The Government of Pakistan has taken various steps to empower women in the country. The major accomplishment in this regard is the increase in women's representation in its all policy-making and legislative institutions. Currently, under the policy of reservation of seats for women, they hold 18 percent of seats in the Senate, 21 percent of seats in the National Assembly, 17 percent of seats in provincial assemblies and 33 percent of seats in local government – figures that are three times higher than those under the previous reservation policy (Mumtaz and Salway 2005). Government's achievements till date also include the formulation of National Plan of Action in 1998, followed by a National Policy for Development and Empowerment of Women in 2002 and Gender Reforms Action Plan in 2004. All these policy documents have been brought out as follow-up actions on the principles enshrined in paragraph 297 of the Platform for Action, adopted by the Government of Pakistan at the Fourth World Conference on Women held in Beijing in September 1995. However, despite commitments to women's empowerment and significant gains on many fronts, water management still remains solely a male preserve in Pakistan. At the same time, women's participation is extremely crucial because of the following distinguished factors that clearly call for greater participation of women at all levels of water management.

Women's Agricultural Contribution: The World Bank's Country Gender Assessment Report on Pakistan (2005a) reveals that in rural Sindh, 33 percent women participate in labor market activities; this proportion of women in Sindh in agricultural labor (that constitutes the highest proportion of economic activities in the country) is higher than that in other provinces. Moreover, 76.9 percent of active rural female labor force is predominantly involved in the agricultural sector. But the activities in which these women are involved are those that offer lower wages, such as picking cotton during the harvesting season. What is also noteworthy is that majority of women, upto 80 percent, reportedly work within their villages. This suggests that employers within the village are likely to face little competition from women workers from outside. It is also apparent that women involved in agriculture are mainly poor, landless, low-wage laborers with limited formal education and mobility. Therefore, such women require greater participation, support and recognition in water management decision-making process that claims to bring about greater equity and equality at all levels.

Women's Land Rights: Women in Pakistan have legal right to inherent family wealth, and yet they rarely exercise this right. A scholar of gender and property law in Pakistan, Rubya Mehdi (2002) points out that although both Islamic and Pakistani state law allows women to inherit immovable and movable property, the tribal nature of social organization in rural Pakistan undermines these inheritance rights (World Bank 2005a: 19). For instance, in Sindh, 95 percent of potential female heirs are those who had no recognized inheritance right or those who had inheritance right but relinquished them (in a vast majority of cases to brothers). Less than 5 percent of women are those, who inherited and either kept the land or sold it (ibid.: 21). Therefore, it may be safe to conclude that the most productive resource, i.e., land, remains under the control of men, a phenomenon that has direct impact on women's right to water because in Sindh, the legitimate right to canal water is confined to land: only a landowner can claim his/her right to canal water. In such circumstances, it is important to enhance women's control over existing assets, particularly land, and thereby facilitate their greater access to water. As Meinzen-Dick and Zwarteveen (1998: 339) observe, "women's participation in organisations with control over natural resources is more challenging (literally) because it deals with property rights over existing resources."

Sustainable irrigation infrastructure: There is significant evidence for the direct and indirect use of irrigation water by women for various purposes such as drinking, washing, bathing, livestock watering, etc. It is widely observed that the gender issues are rarely factored into technical and engineering perspectives on water management. Reeds and Coates (2003) argue that taking an engineer's viewpoint, however, leads to more pragmatic reasons for considering the needs of women. If water is largely used for domestic purposes, it makes sense to ensure that the people most likely to use it are able to use it fully, without any physical, economic and social constraints. Findings from fieldwork also suggest that the current irrigation infrastructure development projects in Sindh have failed to meet the water-related needs and demands of women including children. The third part clearly highlights how engineering structures (washing sites) constructed across water channels, without any consultations with users (mainly women), by highly qualified engineers and consultants have failed to meet the women users' demands. However, women's active participation can lead to sustainable and user-friendly infrastructural development in the province.

All the aforementioned factors strongly justify greater participation and space for women in water management policies and programs. The following section discusses the main drivers of reforms for democratic decentralization of Sindh's irrigation practices, and highlights the claims of popular participation of deprived classes made in the reform program.

Gendered Decentralization of Water Resources

The Indus irrigation system of Pakistan is the largest contiguous irrigation system in the world, and the whole country is extraordinarily dependant on this single river system. At the provincial level, the massive, highly complex and interconnected ecosystem of Indus Basin is managed by Irrigation and Power Department (IPD). Conceptually, the simplest task for water managers in the Indus Basin is to supply water in a predictable, timely manner to those who need it and have the right to enjoy its benefits. Yet, this task is done less and less satisfactorily, lesser during daylight and more behind opaque curtains (World Bank 2005b), resulting in widespread degradation of natural resources, over-exploitation of groundwater, inequitable distribution of water, poor technical performance, drainage and flooding problems, and pervasive environment of mistrust and conflict among provinces, evidence for which is widely available.

In order to respond to all these problems, the Government of Sindh introduced reforms in the management of the entire irrigation system known as democratic decentralization, with the promulgation of Sindh Irrigation and Drainage Authority (SIDA) Act (1997), which was later replaced by Sindh Water Management Ordinance (SWMO) (Government of Sindh 2002). "The form decentralization refers to the transfer of authority for making decisions: to local units of centralized agencies (deconcentration), lower levels of government (devolution) or semi-autonomous authorities (delegation)" (Rondinelli 2002). However, democratic decentralization in Sindh seeks the involvement of users/stakeholders by transferring operational, administrative, fiscal and other powers from government's centralized department to democratically elected and nominated water users' bodies, with the major objective of improving water performance and service delivery. Under democratic decentralization, the following institutions have already been established in Sindh:

(*a*) Sindh Irrigation and Drainage Authority (SIDA),
(*b*) Area Water Board (AWB),
(*c*) Farmers' Organisations (FOs) and
(*d*) Regulatory Authority (RA)

The key elements of the reforms, the new hierarchy of the new institutions is presented in Figure 4.1. The long-term vision is that once the institutions become operational, SIDA would enter into contract with AWBs for bulk of irrigation water and receive effluents generated within the limits of the AWBs. The AWBs would enter into a similar contract with FOs, for bulk supply of water at the head of distributary canals. The FOs would collect the water charges, retain a part of it (40 percent) and pass on the remaining (60 percent) to the AWBs for the maintenance of canals. The AWBs would, in turn, pass on the portion of amount received to the SIDA for the operation and maintenance of the system under

Figure 4.1
Reforms in Sindh Water Management Ordinance (SWMO)

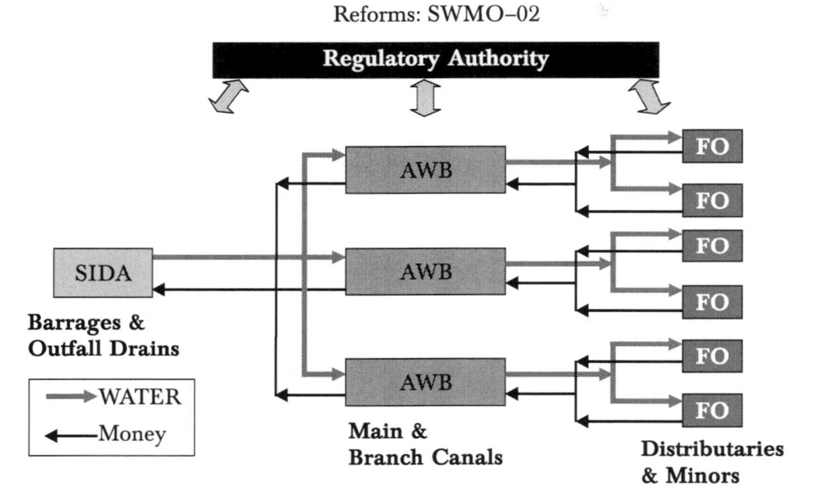

Source: SIDA 2004.

their jurisdiction. It is expected that under the scheme of decentralization, the newly established autonomous, accountable, transparent, cost-effective and customer-oriented institutions would result in improved and sustainable operations through efficient infrastructure leading to higher water delivery, better scheduling, equitable distribution and enhanced agricultural productivity. The Regulatory Authority (RA) is also to be established to promote effective interaction among SIDA, AWBs, FOs including various support bodies.

The decentralization of natural resources in Sindh is justified on the ground that it would ensure full participation and involvement of deprived and marginalized classes of society, for instance, small and tail-end farmers.[1] The SWMO by protecting the rights of the deprived clearly specifies the participation of farmers from head, middle and tail ends, including small farmers, as decision-makers in all decentralized institutions. For instance, the 16-member SIDA board includes five farmers, the AWB consists of 13 members including five farmers, provided that one member shall be a small farmer and one member a farmer from the tail end of a channel.

The FOs also have a significant representation of tail-end and small farmers in their decision-making boards. As Goldfrank (2002: 53) observes, "decentralization and participation are separate but complementary concepts, in that the former is often viewed as a necessary step to achieve the latter." But Meinzen-Dick and Zwarteveen (1998: 337) argue that "devolution of control over resources from the state to local organizations does not necessarily lead to greater participation and empowerment of all stakeholders." This is particularly true of highly differentiated and stratified societies. Sindh is characterized by a highly patriarchal society where inequity not only has a gender dimenson, but a social and economic one. The transfer of power to manage water resources takes place within existing power structures, where women's concerns do not feature. Efforts to control water distribution have already created confrontation between big and small landowners, landowners and tenants (sharecroppers), land users at the head and tail end of the water distribution system. The situation in Sindh parallels that in Nepal: Bienen et al. (1990 cited in Kohl 2003:154) observe that "administrative decentralization in Nepal allowed rich local farmers to capture benefits from locally administrated development projects."

Gender disparities are additional factors which cross-cut all the above, making it sometimes difficult to distinguish between equity issues related to gender and those related to larger power structures. In Sindh, major inequity is experienced by at least half of the population, namely women, who are not actively involved in the reform program at any level (policy, institutional, community or household). The following section shows that despite women's involvement in agriculture, irrigated agriculture and domestic water use, they have failed to participate in water management institutions, are not represented in decision-making bodies and draw no benefits directly generated via the reform program.

Women's Representation and Status at Policy and Institutional Level

Women's representation in water management in Sindh at different levels is alarmingly low. The Irrigation Department has no women professionals working for water distribution. The reform program has also not made any significant progress in bringing about women's participation at the level of policy-making, especially in the decision-making boards, where women have no representation (refer to Table 4.1). However, boards include water management experts like environmentalists, agriculturists, sociologists, farmers and some ex-officio members, all of whom are men.

Table 4.1
Women's representation in decision-making boards of decentralized water management institutions

Organization	*Positions*	*Number of Women*
SIDA Board	Board members	Nil
AWBs	Board members	Nil
FOs	Member in BoM	07 (0.2 percent of total membership)
FOs	Chairperson	Nil

Source: SIDA, June 18, 2008.
Note: BoM – Board of Management.

In 2003, however, under the reform program, few women professionals were employed to carry out field-based activities and office work in the Irrigation Authority. But the gendered division of tasks continues to exist, since senior level positions and engineering positions are only occupied by men. Zwarteveen (2008: 111) also observes similar patterns that "the most managers and engineers, in most water management organisations and agencies in most countries are men."

The FGDs with women water professionals working for reform in Sindh reveal that though women are employed in the water sector, their status remains inferior to that of their male counterparts. An enabling environment that promotes women professionals to use their talent and prove their abilities is absolutely not provided by this sector. Other factors such as non-availability of separate toilets and daycare facility, discriminatory social and organizational attitudes and lack of gender sensitivity within organizations do not create conditions for women to play any productive role in the water sector. As a woman professional working in SIDA remarked, "It took us two-and-a-half years to get a separate toilet allotted for women professionals working in offices, but none of us could actually use that toilet because of its inappropriate location" (interview, Hyderabad, April 2008). Another woman professional informed: "During an interview for a job I was shocked, when SIDA Board's senior member commented on my dress code and asked me not to use cosmetics" (interview, Hyderabad, April 2008).

Women working in offices for the reform program also reported issues like harassment, termination, forced resignation and unnecessary pressure to perform field tasks without providing appropriate facilities such as transportation to the field. At an FGD, a woman professional expressed her helplessness:

> Many times I was chased by him (male fellow engineer); he is now making me scared. I am thinking of withdrawing from this office, because respect is of utmost importance for me and my family. I have complained against him and still waiting for a decision. But I am really not hopeful about any change in his behavior with women workers (FGD, Hyderabad, June 2008).

Regarding transportation facilities she observed: "Women staff members are not provided adequate facilities to perform fieldwork. Many times, I paid from my pocket for fuel to perform field related duties, mostly reimbursed after several months" (ibid.).

It is also noteworthy that though gender is an integral part of the development process, engineers rarely have any formal instructions on the topic. Hence, much is needed to be done to enhance gender sensitivity within organizations. As one of the respondents reflected, "It is difficult to work with engineers, who have no experience of working with women professionals. They want women to behave like men" (FGD, Hyderabad, June 2008).

As far as AWBs are concerned, no female staff has been employed so far. The operation and maintenance of larger canal irrigation systems is a domain in which the actual water distribution happens, and which consists of only male staff: engineers, *abdars*,[2] gatekeepers and canal operators next to the users and others with an interest in water. This study demonstrates that women are not only in minority in the water management institutions, but they are also discriminated against in many ways such as being deprived of basic human amenities like separate toilets. Much is, therefore, needed to be done to provide employment opportunities, full support and an enabling environment to women at all levels, so that women can adequately perform their jobs and play a productive role in water management. Though the fact remains that water sector is highly male dominated with little scope for immediate changes in its male-dominated institutional practices, monopolization of water institutions by men must not be ignored but rather be pointed out in order to bring about changes.

Women's Representation at the Community Level

Stakeholders in water management have limited or no perception of the relevance of women's involvement in community-based water management organizations such as FOs. The process of forming FOs is a classic example of this lacuna. The process was completely gender-insensitive, due to the persistence of preconceived ideas of planners that the farmers to be reckoned with in the formation of FOs are only men, that women farmers would not be allowed by their menfolk to participate in the reform program, and that male farmers would refuse to form FOs in case women are involved. The whole process of FO formation silently and conveniently ignored the right of women farmers to participate in the program. In fact, a few women farmers from Heran Minor in Sanghar district complained in the FGDs: "Nobody has ever asked us to join the FO or to become a member. If they had contacted us, we would have joined the FO" (FGD, Heran Minor, May 2008).

However, experience has shown that women's involvement is easier to bring about at an earlier stage of the formation of FOs, rather than after the completion of the process (Meinzen-Dick and Zwarteveen 1998). The women landowners are normally not included in the FOs, play no role in making decisions concerning the use of land or any revenue resulting from the use of land. Active women's participation in the roles, functions and responsibilities of FOs such as water distribution, revenue collection, conflict resolution, etc., as defined by the SWMO, is not attested. A small number of women who have nominal representation in the FOs normally do not participate, as social restrictions often prevent them from sitting in all-male settings or discussing matters with men. As a married woman farmer from Duthro Minor reflected, "There is no tradition of women going out for such purposes. They have no responsibility in FO affairs or conflicts. Men don't like women to be involved in FO meetings. It is against their honor" (ibid.).

It has also been observed that women, more than men, experience multiple time and mobility constraints due to their multiple responsibilities, in different ways. This has implications for their ability to participate in different activities of FOs, for instance, attending meetings, trainings, etc. Restrictions on their mobility outside home, particularly those arising out of concerns for their security and reputation, are also important factors that determine their own preference for participating in the reform program. Hence, as long as these concerns remain, so will the restrictions. Aa a female participant in the FGDs informed:

> It was not possible for me to be a part of the FO, as I had the responsibilities at home, had to look after children. To be a woman was also a restriction. In villages, women hesitate to go out. The women, who go outside, are not considered respectable. That is why women prefer to be indoors (FGD – Bagi Minor, District Mirpurkhas, May 2008).

No direct benefits for women can be currently seen in the transfer of responsibility for water management from government to users, except an increase in their participation in agricultural labor and their resultant contribution to increased agricultural productivity. But their increased presence in the labor force does not guarantee increase in wages, due to the monopoly and control of local big landowners over agriculture and water resources and lack of any mechanism to organize women to fight against such discrimination. The FOs are also less attractive to women as than are other community organizations such as those formed by non-govenmental organizations (NGOs) that offer them better opportunities for participation. As one of the women farmers explained, "We have Women's Community Organization in our village formed by a local NGO. The NGO offers credit, training, and other forms of funding and opportunities to women. The FO offers nothing to women, it is very much a man-focused organisation" (FGD – Duthro Minor, District Sanghar, June 2008).

However, the reform program in Sindh province has made some progress in the mobilization of women farmers by forming 30 Women Farmers' Groups (WFGs). These groups are formed by women water professionals working in SIDA. The social mobilization of female staff has also led to identification of 2,245 women landowners in the project area. These groups, however, do not perform any active role in the FOs. The experience of these groups suggests that male farmers' mobilization and cooperation is, nevertheless, crucial for the participation of women in the reform program. Women farmers feel more confident when they are fully supported by the male heads of their households. As such, there is not much investment in the WFGs, but women farmers are provided a few introductory trainings under the reform program. Unfortunately, these WFGs are neither recognized by any law like SWMO nor do they have any clear agenda for women's greater participation in water management.

The SWMO is the prime legal and institutional reference point for defining water management in the province. It, however, does not specify gender-specific actions by any agency or organization concerning the devolution of the responsibility for water management. There remains an implicit assumption by the implementers of reforms that all tasks, responsibilities and decisions would be made by men. Thus, they fail to recognize women's different roles, needs and productive contribution towards water management. To a greater extent, the reform program reinforces the existing gender and power relations, as exemplified by the membership criteria of FOs which ensure that only landowners can be their members, simply excluding poor, landless farmers including women who generally do not possess or have any control over land. In this way, potential women farmers' immense contribution to agriculture as paid and family farm laborers is also ignored.

The following section highlights how ignorance or non-recognition of women's water needs and participation has adversely impacted the efficiency and effectiveness of water infrastructure, thereby making a strong case for the argument that the reform program must recognize women as central to water management.

Gendered Aspects of Community Infrastructure

The assumption underlying water policies that irrigation infrastructure is only meant to respond to the farming needs of mostly men is absolutely wrong. There is widespread evidence available showing the use of irrigation water by men, women and children for productive and domestic purposes including livestock bathing. However, in Sindh, irrigation projects and organizations are mainly designed with the notion that the primary use of water is only for agricultural purposes. Fortunately, this efficiency-oriented approach is now being identified as limiting the potential for investments to promote inclusive economic growth because in countries such as Pakistan, many may be excluded from the potential benefits of such programs.

Under the reform program in Sindh, about 100 farmer-managed irrigation channels/distributaries have been rehabilitated. Farmers were fully consulted and involved in the project design and implementation processes by authorities, especially in identifying rehabilitation works. Male farmers fully participated in joint walk-throughs of channels in which they also proposed the construction of multipurpose community infrastructure, i.e., foot bridges, washing and livestock bathing sites, etc. However, it is to be noted that in this whole process, women water users were neither consulted by male farmers nor by any organization. Rather, it was assumed that representatives of male farmer represented the whole community. On the other hand, engineers also failed to recognize the importance of knowing women users' water needs. Even basic drinking water facilities were not included under the purview of rehabilitation works. This was partly because arranging for water for domestic purposes is mainly a responsibility of women and children. Women's invisibility and lack of gender-sensitivity of authorities clearly had a negative affect on mainstreaming women's water needs in the reform program. In the FGDs, women farmers and water users reflected upon washing sites constructed for women in channels and mentioned that the engineering structures developed by highly qualified engineers are not user friendly. As a woman user of a washing site explained: "There is no drainage outlet for the washing sites. The washing activity in the channels increases contamination, because of the use of detergents and in turn cause health hazards, since the water in the channels is also used for drinking by humans and animals" (FGD – Bagi Minor, District Mirpurkhas, May 2008). A woman user of another washing site added: "Engineers have no experience of washing clothes in the channels and, therefore, have no idea of users' demands/needs. A pregnant woman can even lose her fetus as there is no proper place to sit comfortably to wash clothes" (FGD – Duthro Minor, District Sanghar, June 2008).

Women farmers also criticized the authorities for the faulty construction of the channels: "We cannot use the washing *ghat*s when accompanied by children, because there is a life and security threat to them. The channel has no boundary wall . . . children may fall into the channels especially when the channel carries more water" (FGD – Bagi Minor, District Mirpurkhas, May 2008).

All the aforementioned observations by women water users show that water engineers should have an awareness and informed understanding of the needs of all citizens and readiness to address them – that is, an understanding of who they are working for and why. This also raises the question of effectiveness and efficiency of irrigation infrastructure and its impact on potential users' needs. Water management authorities must realize that users' participation is extremely important in planning, designing and construction processes in order to maximize its contribution towards improving the lives of the vulnerable population. In the FGDs, women water users also mentioned that it is less likely to find any community infrastructure, especially in the main canals, for washing, bathing or drinking purposes – the sort of infrastructure that can help users easily access canal water without putting their lives at risk or without contributing to environmental degradation.

It is important to realize that ensuring the responsiveness of water infrastructure to all segments of society would raise the water security of the vulnerable people and contribute towards increasing the effectiveness of the infrastructure itself. Water engineers must realize that their gender-insensitivity disadvantages poor and socially marginalized communities. If irrigation schemes are better planned and designed for multiple uses, in which domestic uses are also given priority, more benefits can be reaped from the same irrigation scheme, especially for women. This may also serve as an entry point for women to participate in water management organizations.

Conclusion

Despite reforms and commitments to water users' participation in water management, women largely still remain invisible in most of the water management policies and institutions. More specifically, it seems that the reform program in Sindh has a long way to go in order to involve women in water management at all levels (policy, institutional and community). Women's effective participation requires support through a legislative framework, and gender equity in water management requires legal recognition and status to mitigate inequality and inefficiency of irrigation policies and programs. As has been pointed out in this essay, the implementation of the reform program, in many cases, has reinforced unequal gender and power relations through formal and informal rules and membership criteria that exclude poor, landless and women from participation in water management schemes and programs.

It is also evident that women use water resources for both domestic and productive purposes, and ignoring their participation and water-related needs may adversely affect the efficiency and effectiveness of water resource management, as exemplified by case of washing sites. Even for such simple engineering designs, there is a vital need to conduct discussions with the user groups, or else the whole exercise would be futile. The case study of Sindh clearly shows that lack of gender sensitivity on the part of the body responsible for implementing the reform program has had a negative impact on the lives of the poor, the landless and women, as also the efficacy of the program itself.

Notes

1. Tail-end farmers are those whose lands are located at the tail of channel. They mostly own lands located in 33 percent of the tail portion of channel/minor.
2. The person who assesses water charges.

References

Agarwal, A. and E. Ostrom. 2001, "Collective Action, Property Rights, and Decentralization in Resources Use in India and Nepal," *Politics and Society*, 29(4): 485–514.

Cleaver, F. T. Franks. 2005, "Water Governance and Poverty: A Framework for Analysis." BCID Research Paper 13, University of Bradford, UK.

Department for International Development (DFID). 2000. "Addressing the Water Crises: Healthier and More Productive Lives for Poor People." Strategies for Achieving the International Development Targets Paper Series, DFID, London, UK.

Fraser, E. 1996. "The Value of Locality," in D. King and G. Stoker (eds), *Rethinking Local Democracy*, pp. 89–110. Basingstoke and London: Macmillan Houndmills.

Gender Water Alliance (GWA) and United Nations Development Programme (UNDP). 2006. *Mainstreaming Gender in Water Management*, Resource Guide, http://www.wsscc.org/sites/default/files/publications/gwa_resource_guide_mainstreaming_gender_in_water_management_2006.pdf (accessed July 3, 2007).

Goldfrank, B. 2002. "The Fragile Flower of Democracy: A Case of Decentralization/Participation in Montevideo," *Politics and Society*, 30(1): 51–83.

Government of Pakistan. 2002. *Pakistan Integrated Household Survey (PIHS) 2001–2002*. Islamabad: Pakistan Bureau of Statistics.

Government of Sindh. 2002. *SWMO (Sindh Water Management Ordinance) 2002*. Karachi: Government of Sindh.

Heller, P. 2001. "Moving the State: The Politics of Democratic Decentralization in Kerala, South Africa and Porto Algere," *Politics and Society*, 29(1): 131–63.

Hellum, A. 2001. "Towards a Human Rights Based Development Approach: The Case of Women in the Water Reforms in Zimbabwe," *Law Social Justice and Global Development Journal*, http://www2.warwick.ac.uk/fac/soc/law/elj/lgd/2001_1/hellum/ (accessed July 23, 2013).

Kohl, B. 2003. "Democratizing Decentralization in Bolivia: The Law of Popular Participation," *Journal of Planning, Education and Research*, 23(2): 153–64.

Koppen, B. Van and I. Hussain. 2007. "Gender and Irrigation: Overview of Issues and Options," *Irrigation and Drainage*, 56(2–3): 289–98.

Koppen, B. Van. 2000. "Gendered Water and Land Rights in Construction: Rice Valley Improvement in Burkina Faso," in B. R. Burns and R. Meinzen-Dick (eds), *Negotiating Water Rights*, 83–104. New Delhi: Sage.

Manor, J. 1999. *The Political Economy of Democratic Decentralization*. Washington DC: World Bank.

Mehdi, R. 2002. *Gender and Property Law in Pakistan; Resources Discourses.* Lahore: Vanguard Books Pvt Ltd.

Meinzen-Dick R. and Margreet Zwarteveen. 1998. "Gendered Participation in Water Management: Issues and Illustrations from Water Users Associations in South Asia," *Agriculture and Human Values*, 15(4): 337–45.

Mumtaz, K. and S. Salway. 2005. "'I Never Go Anywhere': Extricating the Links between Women's Mobility and Uptake of Reproductive Health Services in Pakistan," *Social Science and Medicine*, 60(8): 1751–65.

Reeds, B. and S. Coates. 2003. "Engineering and Gender Issues: Evidence from Low Income Countries," *Municipal Engineering*, 156(ME2): 127–33.

Rondinelli, Dennis. 2002. "What Is Decentralization?," in Jennie Litvack and Jessica Seddon (eds), *Decentralization Briefing Notes*, pp. 1–5. Washington DC: The World Bank Institute.

Sindh Irrigation and Drainage Authority (SIDA). 2004. "Water Reforms in Sindh Presentation," in District Council Thatta, Sindh, 25 April.

Upadhyay, B. 2000. "Poverty, Gender And Water Issues in Integrated Agriculture And Irrigation Institutions: Mainstreaming Gender in Water Resources Management." *Water Policy Research Highlights-4*, IWMI and Tata Water Policy Program.

Walker, M. 2006. "Women, Water Policy and Reform: Global Discursion and Local Realities in Zimbabwe." Working Paper 287, Michigan State University, Michigan.

World Bank. 2005a. *Pakistan Country Gender Assessment: Bridging the Gender Gap: Opportunities and Challenges.* Report 3244-PAK Washington U.S.A., http://www-wds.worldbank.org/external/default/WDSContentServer/WDSP/IB/2006/02/01/000160016_20060201091057/Rendered/PDF/322440PAK.pdf (accessed on January 23, 2008).

———. 2005b. *Pakistan's Water Economy: Running Dry.* Islamabad: World Bank.

———. 2003. *Water Resources Sector Strategy: Strategic Direction for World Bank Engagement.* Washington DC: World Bank.

Zwarteveen, Margreet. 2008. "Men Masculinities and Water Power in Irrigation," *Water Alternatives*, 1(1): 111–30.

5

An Attempt at Quantification of Women's Empowerment in Small-Scale Water Resources Project

Sayeda Asifa Ashrafi and *Rezaur Rahman*

In Bangladesh, women are more vulnerable than men to chronic poverty due to gender inequalities in various social, economic and political institutions. Such inequalities can be found in the andocentric distribution of income, control over property or income and access to productive inputs (such as credits), decision-making processes, and rights and entitlements. Women also have limited access to and control over water resources as they have with respect to all other resources. Water is a social and economic good. On the one hand, it acts as an entry point for sustainable development and poverty reduction. On the other hand, inequality in access to water resources, water-related diseases and disasters proliferate cycles of poverty.

Government of Bangladesh declared National Women Development Policy in 1997 (Government of People's Republic of Bangladesh 1997), which was a step towards ensuring women's equal participation in development. It subsequently declared National Water Policy in 1999 (Government of People's Republic of Bangladesh 1999), highlighting women's participation in water sector. In an attempt to realize this vision, the Local Government Engineering Department (LGED) has given greater importance to gender and development and has been working to promote gender equity through Small Scale Water Resources Development Subprojects (SSWRDSP) (LGED 2003). The project has opened opportunities for women to increase their income and participate in economic activities, but the extent to which the sub-projects have contributed to creating a situation of more equal opportunities for poor men and women cannot be assessed (BUET, BIDS and Delft Hydraulics 2003). The aim of this chapter is, therefore, to quantify women's empowerment from household to society, through their participation in small-scale water projects.

Study Area

For this study, three sub-projects from the first phase of SSWRDSP Brajamul-Bhitikhal Flood Control and Drainage (FCD) – (*a*) sub-project of Saorail union of Pangsa thana, Beel Salua Drainage & Water Conservation (DR & WCS); (*b*) subproject of Baharpur union of Baliakandi thana and Bara Nurpur Water conservation (WCS); and (*c*) subproject of Shahid Wahabpur union of *sadar* thana of Rajbari district – were selected purposively. For the purpose of comparison of women's empowerment through participation in water management, a control village, Char Tetulia (without project), with similar geographical and demographical characteristics was also selected.

Emergence of a gender dimension in water resources management is very recent in Bangladesh. However, the participation of stakeholders, especially women in the development process, is by no means emphasized. Various studies have been done in the areas of women in development, and gender and development for water resources management (GWA 2005; Parpart et al. 2000; Government of the People's Republic of Bangladesh 1999; Visvanathan et al. 1997; Batliwala 1995; Duza and Begum 1993; Rathgeber 1990; Parpart 1989; Sen and Grown 1987). More often, women are seen as participants in water supply and sanitation projects (ADB 2006a; Kabir and Faisal 2005; Nahar 2002; Ahmed and Jahan 2000). Nowadays, women's participation is a priority component of project management under integrated water resources approach. Women's participation in such development process and their empowerment as an end result is a very important issue, as has been pointed by various studies (Brandi 2005; Chowdhury 2005; Biswas 2004; Goswami 1998; Hashemi et al. 1996; UNDP 1994; Ward 1992; Chowdhury 1990). However, no study on the

Map 5.1
The study area in Rajbari district, Bangladesh

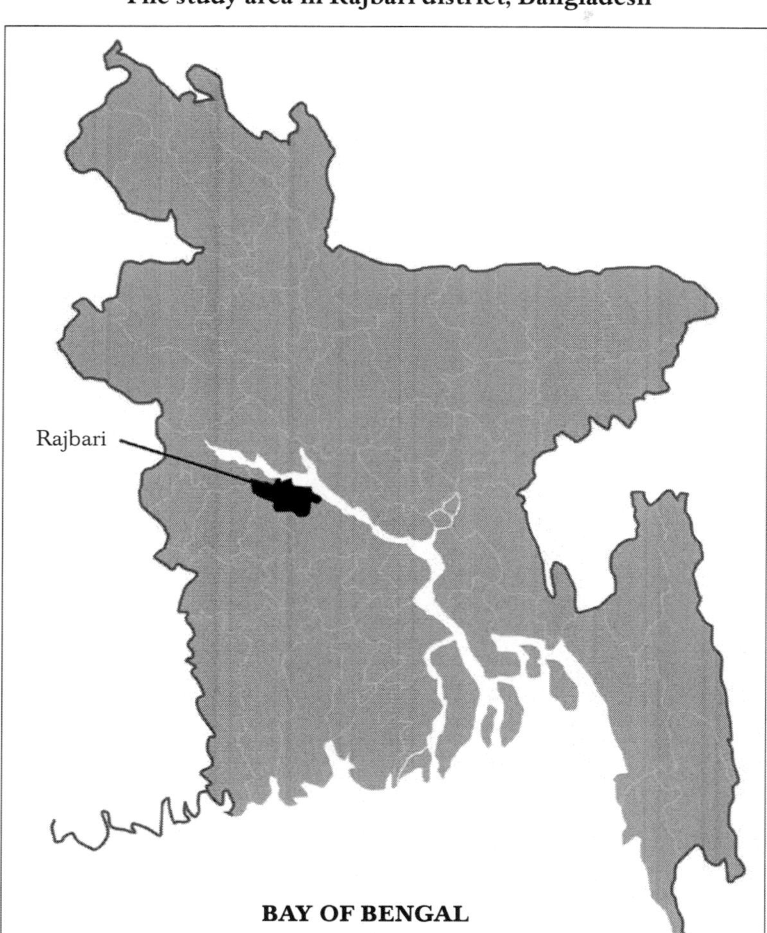

Source: Prepared by the authors. Map not to scale.

estimation of real changes in women's lives due to their participation in water resource management has been made; therefore, the present study is an attempt to quantify women's empowerment through their participation in water resource management.

Methodology

The present study is concerned with women's participation in water resource management. The study area was selected in such a way that all dimensions of water resource management like, flood control, drainage, water conservation for irrigation and domestic usages, etc., could be fully covered. Therefore, different types of project sites of SSWRDSP under each category were selected as the study area.

Woman as an individual was considered the unit of analysis in this study. For the collection of primary data, systematic random sampling of the members of Bangladesh's Water Management Cooperative Association (WMCA) was done. To avoid any bias, the first respondent was selected randomly from the list of beneficiaries, every 10th sample was taken and thus 10 percent of the total sample was covered. The total number of WMCA members was around 1,600, of which 40 women from the study area were selected for structured interviews and around 120 men and women were selected for focus group discussions (FGDs).

The primary data was collected from the field in two phases. First, structured interviews were conducted in the study area to obtain individual responses of participants to different issues regarding empowerment. A certain weightage was assigned to each response in order to calculate the qualitative data quantitatively. Some parts of the questionnaire were kept open-ended to serve the pertinent functioning of the study. Second, FGDs were conducted to collect qualitative information and thereby understand the factors affecting women's participation in water resource management, as also identify newer areas of activity that the rural community thought women could participate in. The focus group comprised of male and female WUA members, WUA chairperson, farmers, fisherman, fish farmers, local Union Parishad members and chairman. Two types of FGDs, i.e., with a women's group and with a mixed group of women and men, were conducted in order to comprehensively analyze both men's and women's perspectives on equal participation in water management. In each sub-project area, two FGDs were conducted with the aforementioned participants.

The data collected from the field study was analyzed in two different stages: in one stage the data was assigned unequal weightage, and in another, it was given equal weightage. For assigning weight, six eminent experts working in water sector in different areas, e.g., policy, planning and operation, etc., and in different capacities were selected by the authors of this study in consultation with professors from the Institute of Flood and Water Management, Dhaka. Among the six experts, two were from LGED and one each from Action Aid Bangladesh, a non-governmental organization (NGO); Bangladesh Water Development Board; United Nations Children's Fund (UNICEF); and Bangladesh Water Partnership. The researcher collected this information through direct discussions with the respondents. On the basis of the responses of the respondents, the weighting system was developed.

The prior studies/researches on the issue of women's role in water management have hitherto been focused exclusively on the responsibility of women for the provision and management of water at the household level, and less on their active participation in planning, design and management, while their real gains from their participation were not studied. This study seeks to break new ground by quantifying the level of changes in women's lives as a result of their participation in water resource management through some indicators. In this study, women's empowerment has been measured on the basis of metrics devised by the author and the Empowerment Index (Biswas 2004, 1999). In the existing literature, empowerment was mostly based on social and demographic dimensions and the issues are not stated clearly. In this study, selection of issues is based on literature review (Biswas 2004, 1999), and also by a general observation of women's involvement in water-related activities and WUAs and consultation with the experts in this field.

This study considers the issues of women's mobility, decision-making power, autonomy, economic status, exposure to information and institutional involvement. Indicators for these issues were selected in the same way as the issues themselves were. In addition, these indicators were also chosen in consultation with experts who have been working in the area of water and gender in Bangladesh and were especially selected for this study.

Measuring Women's Empowerment

The empowerment index has been measured in two stages. In the first stage, empowerment index of a woman for a single issue consisting of different indicators is developed. In the second stage, a composite empowerment index for a woman consisting of different issues is developed in order to assign equal and unequal weights. In fact, the different indicators considered under each issue are not equally important in relation to women's empowerment. Consideration of equal weight for all the individual indicators of an issue leads to the possibility of some or the other error creeping into the calculation of composite index. Therefore, in order to ensure the accuracy of the estimate, a weighting system has been followed.

Empowerment Index of a Woman for Single Issues of Different Indicators with Equal Weights

In the first stage, the empowerment index of a woman for a single issue with different indicators can be calculated by the following formula:

$$EI_{ij} = \frac{X_1 + X_2 + \dots\dots\dots\dots\dots + X_n}{M} \times 100$$

$$= \frac{\sum_{j=1}^{n} X_j}{M} \times 100$$

Where,
EI_{ij} = Empowerment index of ith woman for j^{th} issue.
X_j = Value of individual indicators of jth issue.
M = Maximum possible score or outcome.
= Number of individual indicator of an issue multiplied by the maximum score assigned for individual indicator of that issue.
n = Number of individual indicators of an issue.
The maximum value of empowerment index will be 100 according to this formula.

In the second stage, the composite empowerment index of a woman consisting of different issues with equal weight can be calculated by the following formula:

$$EI_i = \frac{EI_{i1} + EI_{i2} + \dots\dots\dots\dots\dots\dots + EI_{iN}}{N}$$

$$= \frac{\sum_{j=1}^{N} EI_{ij}}{N}$$

Where,
EI_i = Composite empowerment index of a woman.
N = Number of issues considered in the composite index.
The highest possible value of composite empowerment index will be 100 according to this formula.

Empowerment Index of a Woman for Single Issue of Different Indicators with Unequal Weights

Opinion Survey Method: Under this method, experts who are professionally involved in empowerment issues were consulted on all the indicators for different issues and asked to assign a score to each indicator on a scale from 1 to 5 on the basis of the degree of relationship of the issues with the empowerment level. The higher the degree of relationship of an indicator with the empowerment issue, the higher was the score assigned. Then the average weights for the indicators were calculated within 100 by the following formula:

$$W_i = \frac{X_1 f_1 + X_2 f_2 + \dots\dots\dots\dots\dots\dots + X_n f_n}{M} \times 100$$

$$= \frac{\sum_{i=1}^{n} X_i f_i}{6 \sum_{i=1}^{6} f_i} \times 100$$

Where
W_i = Weight of an individual indicator of each issue.
X_i = Individual score assigned by the respondents on a scale from 1 to 5.
f_i = Frequency of responses of individual score.

A checklist consisting of different empowerment indicators for each issue with a score on a scale from 1 to 5 was designed and six experts, both women and men representing government organizations, international organizations and NGOs were asked to assign score to each issue.

In the first stage, the weighted empowerment index of a woman for an issue consisting of different indicators can be calculated by following formula:

$$EI_{ij} = \frac{W_1X_1 + W_2X_2 + \dots\dots\dots\dots + W_nX_n}{MS(W_1 + W_2 + \dots\dots\dots\dots + W_n)} \times 100$$

$$= \frac{\sum_{i=1}^{n} W_iX_i}{MS\left(\sum_{i=1}^{n} W_i\right)} \times 100$$

Where,
EI_{ij} = Weighted empowerment index of ith woman for j^{th} issue.
W_i = Weight of an individual indicator of each issue derived from expert opinion through survey method.
X_i = Value of response of ith indicator of an issue.
MS = Maximum score assigned for a response of ith indicator of an issue.

In the second stage, the composite empowerment index consisting of different issues can be calculated by the following formula:

$$EI_i = \frac{w_1EI_{i1} + w_2EI_{i2} + \dots\dots\dots\dots + w_NEI_{iN}}{w_1 + w_2 + \dots\dots\dots\dots + w_N}$$

$$= \frac{\sum_{j=1}^{N} w_j EI_{ij}}{\sum_{j=1}^{N} w_j}$$

Where,
EI_i = Composite empowerment index of ith woman.
EI_{ij} = Empowerment index of ith woman for j^{th} issue.
w_j = Weight of jth issue derived by taking weighted average of weights of an individual indicator.
N = Number of issues considered under the composite empowerment index.
Maximum EI_i will be 100 according to this formula.

Empowerment vs Participation

In this study, women's empowerment is seen as a positive trait. The empowerment they have achieved through participation in the water resource management project can be in the form of skill training, increased capacity, access to and control over resources, welfare, etc. These are some of the components that can be used to measure the level of gender equality. Women's empowerment and gender equality have been seen in terms of their functional relationship with these components.

In this section, the level of women's empowerment has been measured through a Women's Empowerment Index (WEI) considering some indicators. Apart from this, an index regression analysis has been undertaken to understand the factors influencing the index. WEI has been developed in two stages: first, as Empowerment Index (EI) for respective issues and second, as Composite Empowerment Index (CEI) for a woman. CEI has been measured in two different ways: first, equal weight has been assigned to each of the issues, and second, unequal weight has been assigned to each of the issues and then CEI has been measured. The study reveals that women in project areas are more empowered than in areas without external interventions. Differences in EI have also been observed among the sub-project areas.

Table 5.1 shows that CEI is the highest in the sub-project (Baranurpur and Brajamul-Bhitikhal) having the highest level of women's participation in WMCA. Among the sub-projects, Beel Salua has the lowest level of average CEI worked out by both methods.

Table 5.1
Average Composite Empowerment Index (CEI)

Composite Empowerment Index (CEI)	*Brajamul-Bhitikhal (FCD Sub-project)*	*Beel Salua (DR & WCS Sub-project)*	*Baranurpur (WCS Sub-project)*	*Control Site (No project)*
Equal weight CEI	68	46	67	34
Unequal Weight CEI	68	47	68	34

Source: Field survey (2006–07).

Table 5.2 shows that among the four study areas, the highest (30 percent of the total) number of women belong to two empowerment levels 31–50 and 51–70, followed by 25 percent in 71–90 category. In the category of 31–50, the average CEI of the control site was around 34 which makes this category significant. Under the category of 51–70, Barajamul-Bhitikhal and Baranurpur

Table 5.2
Area-wise percentage of the respondents under different categories of weighted CEI

	Percentage of Respondent's Empowerment Range				
Project Area	*≤30*	*31–50*	*51–70*	*71–90*	*≥90*
Brajamul-Bhitikhal (FCD Sub-project, n = 13)	0	0	18	15	0
Beel Salua (DR & WCS Sub-project, n = 4)	0	10	0	0	0
Baranurpur (WCS Sub-project, n = 13)	0	5	15	7	5
Control Site (No project n = 10)	10	15	0	0	0
Total (n = 40)	10	30	30	25	5

Source: Field survey (2006–07).

sub-projects have the highest percentage of women (18 and 15, respectively). In Baranurpur sub-project, women within the range 71–90 and >90 is 7 percent and 5 percent, respectively. The study reveals that percentage of women having higher levels of empowerment is more in areas having higher levels of women's participation in water management.

Empowerment Index of Different Issues

Mobility

To assess mobility of women, some of indicators that have been used are: whether they can go alone to places outside villages for attending training in income-generating activities (IGA) or women's meetings, for work in labor contracting society, or for fetching water; to health care centers; to markets; to NGO offices, etc. In most cases, it has been observed that women's mobility is limited within the villages. But they are moving freely to health care centers, to NGO offices and outside villages for fetching water. In Brajamul-Bhitikhal and Baranurpur sub-project areas, all women are involved in the WMCA micro-credit program, as also in banks, insurance companies or development organizations. Women of these areas have also been receiving IGA training from LGED and regularly participating in various WMCA meetings that enable them to go out. They frequently travel to other villages, but in groups. They also work with their husbands in fields, but this mobility is limited to women of small farmer households. With the increase of crop production, decrease in labor supply and high wages, most women from marginal farmer households work in the field. The study reveals that average index of women's mobility is the highest (80) in Brajamul-Bhitikhal sub-project area, followed by Baranurpur (76) (Table 5.3). It has also been noticed that women in Barajamul sub-project area are occupying a superior position in WMCA., and that there is little difference between the Beel Salua and Control site, with Beel Salua being a little on the higher side.

Table 5.3
Project-wise average empowerment index of individual issues

Issues	*Brajamul-Bhitikhal (FCD Sub-project)*	*Beel Salua (DR & WCS Sub-project)*	*Baranurpur (WCS Sub-project)*	*Control Site (No project)*
Mobility	80	38	76	36
Decision-making Power	64	45	72	30
Autonomy	56	39	53	26
Economic Status	80	63	74	46
Exposure to information	71	54	77	46
Institutional involvement	59	40	55	21

Source: Field survey (2006–07).

Decision-making Power

Decision-making power is considered as one of the most influential parameters for measuring women's empowerment. However, women in Bangladeshi rural society, compared to men, enjoy this power in a very limited way.

In most cases, women take some household-level decisions jointly with their husbands or members of WMCA. Only in a few cases like child rearing and visits to natal homes can they take decisions independently. In the sub-project area, women have the power to decide on how to spend their own imcome from wage labor and household income. They are quite independent in taking decisions on water use and involvement with different developmental organizations or cooperative associations. But they take joint decisions, with their husbands, with regard to water-management-related issues like logistics of project implementation, operation and maintenance (O&M) committee meetings, irrigation facilities, etc. This is so probably due to the organogram of WMCA. Only a few female respondents in our study alleged that their decisions are overridden by the male members of WMCA. In the control area, the picture is just the opposite of the project areas. Women there have very limited decision-making powers, something very common in our traditional society. They also have to take their husbands' consent a decision jointly to go to their natal homes, or spend their own income. However, in both areas, women have very little power to take decisions on buying and selling property. The average score assigned to women's decision-making power in sub-project areas is higher than that in the control area, as shown by the values of 72, 64 and 45 assigned to Baranurpur, Brajamul-Bhitikhal and Beel Salua sub-project areas respectively, as against 30 assigned to the control site (Table 5.3). Regardless of other facts, this explains very clearly the functional relationship between decision-making power and women's participation in SSWRDSP.

Autonomy

Women's autonomy is determined by social customs or traditions (Biswas 2004). In a patriarchal society like that of Bangladesh, women are forced to be dependent on their menfolk. Autonomy is very much related to decision-making power; the higher the decision-making power, the greater the autonomy (ibid.). Our study also found that women having a high degree of decision-making power are also enjoy more autonomy. In the project areas where women's participation is high, exercise of more autonomy by them is noticeable. The average index for autonomy issue is the highest in Brajamul-Bhitikhal sub-project area (56), followed by Baranurpur (53) (see Table 5.3). It is high due to a few educated women being involved in the project work and being quite independent of their family members or spouses. They also play a significant decision-making role in the activities of WMCA, e.g., O&M of sluice gates, earth work to be done around canals, etc. In general, most women can talk to unknown persons and cast their votes during election (for both project and control areas). More specifically, women in sub-project areas enjoy greater economic autonomy; a few have access to a share in fisheries and agriculture as well. However, average autonomy index is lower than other issues because of the traditional nature of the society. Even in project areas women have no or little autonomy in family planning, providing economic support to relatives of their natal families and buying household assets.

Economic Status

The issue of economic status has various dimensions. Only a few indicators of women's economic status have been considered in this study like right to land; access to and control over their own income, household income, savings and income from participation in water management projects, etc. One of the main indicators of economic status is the ownership of land. Land is mostly owned by men; women are deprived of this right. Land provides social status, economic security and the right to demand water for irrigation. Around 5 percent of the total surveyed sample population of women in the three sub-project areas and one control site has the right to own land. Access to land enables women to more independently use household income and savings. In course of their involvement in water management projects, women have received training in various types of IGA. Besides, 20 percent of the project earthwork (embankment construction and repair, canal excavation, etc.) is carried out by women in the project areas (LGED 2003). Through this work, the income of women in project areas has increased significantly. Table 5.3 shows that the average index for the issue of economic status is highest in the project area. Among the project areas, change in percentage is probably due to the level of women's participation and education (two women in the Barajamul-Bhitikhal project are are graduates and pursue occupations rather than remaining as full-time homemakers). Unlike women in the project areas, women in the control area have fewer opportunities to earn money and mostly depend on their husbands' income.

Exposure to Information

Exposure to information among women in the study areas varies significantly with respect to reading newspapers and participating in public meetings, workshops and seminars. However, activities like listening to radio and watching television is observed in all areas. Average index for exposure to information is highest in Baranurpur (77) followed by Brajamul-Bhitikhal project area (71).

Literacy rate and presence of WMCA makes a significant difference in other two areas, viz., Beel Salua and control site where this index is 54 and 46 respectively.

Institutional involvement

Institutional involvement varies significantly from place to place. It is observed to be the highest in Brajamul-Bhitikhal area (59), followed by Baranurpur area (55) and Beel Salua (40); and only half of that in Beel Salua is in the control site. The study reveals that in control site the women's institutional involvement is less than that in other areas. This ultimately limits women's empowerment.

Various factors may affect CEI, of which socio-economic factors have been discussed in this section. Understanding this factor is very important for formulating the future development initiatives for women's empowerment. Factors like husband's occupation, total landholding size of the household, study location, income (personal and household), respondent's own occupation, landownership (personal and household), respondent's level of education, involvement with other NGOs and WMCA membership – all significantly influence women's CEI.

Multiple regression models have been run using the method of Ordinary Least Square (OLS). In the initial stage, the aforementioned eight socio-economic factors are considered. But while analyzing the individual areas, due to data limitation of a statistical program, few factors are not considered. For example, in case of control site database on WMCA membership, respondent's occupation, etc., have not been considered. The OLS estimates of the parameters of multiple regression model with their level of significance, adjusted coefficient of multiple determination (R^2) and F-value with their significance levels are presented in the Table 5.5. The fitness of the model is evident from the high value of F as also from the adjusted R^2 value. The adjusted value of R^2 implies that the explanatory variables considered in the model explained up to 62 percent of the variation in women's empowerment in case of combined areas.

Table 5.4 shows that the estimated adjusted R^2 value and F-value of the regression is high in the case of Model II. Therefore,

Table 5.4
Multiple regression coefficient of contributory factors for women's empowerment in the combined areas

Independent Variable	*Variable (CEI)*	
	*Model I**	*Model II**
Constant	36.6804	36.2997
Study location	25.2627	26.0038
Husband's income	–0.0002	–0.0001
Respondent's own/personal income	0.0006	0.0006
Total landholding size of the household	0.0108	–0.0062
Respondent's own occupation	18.5663	18.0359
Respondent's level of education	4.2288	4.4979
Involvement with NGOs/WMCA micro-credit program	6.7074	6.4759
WMCA membership status	–1.5219	–1.2035
Adjusted R^2	0.6292	0.6298
F-value	8.3516	8.3734

Source: Field survey (2006–07).
Notes: Model I is for equal weight method, Model II is for unequal weighted method.
* *The* model has been run at 5 percent level of significance.

Table 5.5
Reasons cited by respondents for women's participation in water management

Reason	*Percentage of Respondents Citing the Reasons*
To get water properly round the year	11
To be economically self-sufficient	4
To mobilize/lead other women	6
To know more about water management	15
To work for the locality	6
For savings	32
To gain right to access and use water	4
For recreation	4
To meet with outsiders	4
To earn money	11
To get training	2
Total	100

Source: Field survey (2006–07).

Model II may be preferred in terms of applicability. In the case of Model I, each of the indicators is not equally important for the respective issue. On the other hand, while eliciting expert opinion, experts were requested to assign a score to each of the indicators on a scale from 1 to 5 on the basis of the degree of relationship of the indicator with the empowerment level especially related to water management.

Study Location

Though the four locations are in the same district, there are significant differences between the villages with respect to water projects, level of education and cultural practices that influence women's empowerment. It has been observed that on an average, the empowerment level of women in project areas is over 26 percent higher than that of women in non-project areas.

Occupation and Income (Personal and Household)

Most of the women in both project and non-project areas are economically dependent on their spouses. This dependence acts as a constraint for their empowerment. High economic status is ensured when woman can earn money for herself and also has access to and control over that income. When a woman has the mobility to pursue an income-generating occupation, other than her usual mobility for performing daily chores, her exposure to information automatically increases. Such economic empowerment then provides security to women. Economically secure women feel free to go out. Thus, both woman's own and household income significantly influence the degree of her empowerment. It can be observed from Table 5.4 that the respondent's own/personal income has a more positive impact on the level of her empowerment than does household income. The increase in respondent's own income by a single unit is seen as leading to 0.06 percent increase in CEI. Similarly, respondent's own occupation has significant influence on her level of empowerment. It is observed that 18 percent of the change in CEI owes to change in occupation.

Total Landholding Size of the Household

Rural women in a country like Bangladesh seldom enjoy the right to own land. Though some women have their own land, they have no access to and control over it. Total household landholding size has a significant influence on CEI. It is observed that with the decrease in the amount of land owned by the household, a woman's empowerment increases. Another percentage analysis also shows that the wives of marginal and small farmers "greater mobility and freedom than do those of medium farmers." As the women in smallholding farmers' households, like their spouses, have to go outside the village for work, they enjoy more autonomy and mobility. As they work independently in fields, they can also take some decisions, such as purchasing household goods, grocery, etc. A woman in a medium farmer's household, on the other hand, sometimes, only takes decisions for her children's well-being, but hardly takes part in household income-generation and consequently has less access to and control over the household income and resources than does a woman of a smallholding famer household. Thus, compared to a woman in a medium farmer's household, a woman in smallholding farmer's household is more empowered, even though the household income of the medium farmer's family may be higher.

Respondent's Level of Education

Level of education is a major factor for women's empowerment. Education enables women to learn what to do or what not to do. It also provides them opportunities; educated women can challenge their traditional roles in a society. Education is very much related to income, occupation, exposure to information, etc. Our study reveals that women's level of education has a significant influence on their level of empowerment: 4 percent change in CEI is observed due to change in the level of education by a single unit.

Women at Water Work

The economy of Bangladesh is predominantly based on agriculture. At the field level, gender division of labor is less among the small farmers. In such cases, women work in the fields with their husbands. There are also good examples of women's participation in water supply and sanitation systems in Bangladesh, especially in rural areas. Thus, women's access to resources is an emerging concern. Apart from surface water management, the concept of water management has been extended to include watershed management, e.g., water for all living beings, environment, agriculture, domestic use, etc. So, women should take part in water management from this extended conceptual viewpoint wherein the roles and responsibilities of a woman should be reconsidered.

Women are now being considered important stakeholders in water management by the donor agencies and the Government of Bangladesh. Women have responsibilities as mothers, homemakers and leaders in a society; when these roles blend well, total development is achieved. Women leadership development in water management is of great importance, as women are the main users of water. Findings from focus group discussions (FGDs) show that most respondents thought that women's participation is needed for efficient water management. The opinions of participants in the FGDs are summarized below.

Participation in water management helps women pursue income-generating activities, so they should take part in water management. In most of the cases, women always stay at home so they can operate the sluice gate, pump or nearby irrigation canal more readily than a man can who is always busy with in other economic activities. Women are more capable of mobilizing people than are men in a locality, so they can contribute differently and significantly in water management. Participation in water management can act as a motivating force for the women to come out out of their traditional lifestyle. For women, savings have the highest priority (32 percent) followed by acquisition of knowledge about water management. Eleven percent women opined that they should engage in water management in order to get water for various purposes round the year. Such responses suggest that women in the sub-project areas are now concerned about and thinking of their involvement in water management in their own interest; they want to feel more secure, empowered not only in the family but in the society at large.

Our study also reveals that 50 percent of the women interviewed thought that they could directly contribute to water management; of this, 27 percent were from Barajamul-Bhitikhal and 20 percent from Baranurpur sub-project area. Data show that in areas where women are more empowered due to the project in operation, they are also conscious about their roles and responsibilities in water management. It also suggests that empowerment and women's participation in water management have a functional interrelationship. Women's involvement in water management has emerged in the last few decades, but that was mostly limited to water supply and sanitation (WATSAN).

Participation and Changes

Benefits accruing to women due to their participation in water management projects have been, as cited by the respondents, at four levels: society, family, personal and others. The impact at the societal level goes beyond the individual woman. Most of the responses highlighted economic change in the society due to project interventions. About 45 percent of the respondents were women who had benefitted from increased cropping intensity and huge crop production due to proper irrigation facilities. This increased production had a positive impact on the family income of the women as well.

Due to the agro-based character of the economy, most of the women in rural areas measure their social status and happiness in terms of increased production. Nineteen percent of the responses pointed out reduced poverty in the areas due to increased crop production. Thirteen percent of the responses suggested that distressed women in the project areas have been employed due to the projects. Besides, income-generating activities, e.g., O&M

work for sluice gates, embankments, labor contracting for earth works, etc., for women have increased in general, thereby making them economically empowered in their respective villages. Family-level benefits also accrued to individual women: 48 percent of the responses pointed out direct benefits to their families due to their participation. Due to their participation in water management projects, women could easily and efficiently use water for domestic purposes. Their difficulties in accessing water for domestic use and planning its domestic use have decreased. Most of the families' annual income and savings have also been increased positively impacting their living standards. Due to their participation in WMCA, women have had an opportunity to meet and share their experiences, knowledge and perceptions on different issues.

Women's knowledge on water management issues has increased due to project interventions. Further, not only their participation in water management but also their leadership capacity has increased in their respective localities. Their awareness of women's rights and violence against women has increased as well. They feel themselves to be more dignified as they can take their decisions in the family and society. This learning from their participation in water management can easily be transmitted to the next generation.

Table 5.6 shows that 40 percent of the respondents – 20 percent of the total sample group responded and the rest had no comments on this issue – think that women in the project areas are getting encouraged to see themselves as part of the WMCA. Such women are more capable of making people understand the significance and potential of water management. Gender issues in water management are getting prioritized; women's needs, choices and voices are being factored into designing water projects; women's access to and control over water resources are being ensured through gender equality in water mnagement projects from designing to implementation stage.

If women are left out of the ambit of water management, then around 50 percent of the total population would be deprived of the benefits of water resource development. So, women's participation is essential for efficient water management. The present state of women's participation alerts us to the need for their greater

Table 5.6
Women's contribution to society as participants in WMCA

Comments	*Percentage*
Capability to make people understand about water management has developed.	10
Support in favor of distressed women has become possible.	20
Other women are getting curious about and encouraged to participate in water management.	40
Gender issues are being prioritized in meetings	20
Forms of violence and injustice against women (e.g., dowry, divorce, early marriage, etc.) have decreased.	10
Total	100

Source: Field survey (2006–07).

participation: the extent of their current involvement is still not very satisfactory and needs improvement. Keeping this aspect in mind, an attempt has been made to explore some ideas regarding opening up newer areas of their involvement in the water sector. From Table 5.7, it is evident that women are more interested in irrigation water management in future especially for farming in home and outside. Around 15 percent women responded in favor of their participation in water management for irrigation. Women in FCD and WCS & DR sub-project areas mentioned that they could maintain or regulate the pumps, irrigation channels, etc.

Table 5.7
Percentage of responses pointing to newer areas of women's participation in water management

Response	*Percentage*
Building awareness on the need for safe/efficient water use by women	10
Mass mobilization	12
Irrigation work and domestic water supply	15
Community fishing in canal or *bheel*	10
Large-scale vegetable production at home and on embankments	13
Duck rearing	7
More focus on O&M work	9
No comments	22
Total	100

Source: Field survey (2006–07).

Landownership, however, still acts as a major constraint to their direct involvement in irrigation water management, since most women do not own land. Small farmers, however, always work on others' croplands as day laborers other than their cultivating their own small plots. Since the small plots are mostly cultivated by farmers and their wives, the latter have to directly participate in both household work and work in the fields including watering of crops. Even though crops are watered through pumps, women can play a vital role in irrigation water management. In some areas, water meant for irrigation is put to domestic use during periods of water scarcity. In this way women manage to deal with water deficit.

Women leaders can play a vital role in social mobilization than can men. Women can, by way of mass mobilization, help promote small-scale cropping, seed preservation, water pollution control, safe or efficient water use by the community, etc. Women in the project areas have a great opportunity to engage in and promote aquaculture. Most women in WCS project area opined that they could go for community fish production with their menfolk. Women in FCD project area also agreed that they could cultivate fish in canals or ponds. Besides, Women in project areas can go for large-scale duck rearing.

Apart from water management, women can intensively take part in O&M work for water-related infrastructure, control the water level around the sluice gates and clean the gates. Women's responsibilities can be extended to collection of monetary contributions from the community for O&M, as they are capable of and can prove efficient in this work. Generally, total household management depends on efficient water management, so women's direct and indirect participation in water management has a great importance in integrated water resources management (IWRM) as well.

Women's participation in water management may be limited by various factors. In this study, it is discussed in two stages; first, factors limiting their efficient working in WMCA, and second, factors hindering their participation in the project. Around 63 percent of the respondents confessed that they cannot work efficiently in WMCA, while 33 percent revealed that they face problems in their participation in the project. In the rural society, the level of women's technical knowledge on water management is still quite poor. This also hinders their active participation in the water sector. Their menfolk can, on the other hand, meet a number of people and collect relevant information as also enjoy other facilities by virtue of their far greater mobility and freedom. This creates a significant knowledge gap between men and women. Women work more within the households, a situation that makes their work remain invisible and unrecognized. Greater workload at the household level makes it difficult for them to involve themselves in any other activity. Besides, non-cooperative attitude of men towards women members can account for their poor performance in WMCA. Women in project areas have identified some factors that limit their participation in SSWRDSP. These include cultural barriers and lack of awareness that have prevented them from coming out of the confines of their households for development activities in water management. Family restrictions on going outside the village is a common problem faced by most women, especially those of big farmers' households whose male heads are less interested in encouraging their participation in water management projects.

Way Forward

Capacity-building at grassroots and national level and minimum level of technical knowledge on water management are the keys to women's active participation in water management. At the same time, conducting training programs on capacity-building in the areas of gender equality and IWRM are very essential for planning and implementing agencies as well. Mainstreaming gender in water management: Our studied projects do not explicitly aim to ensure women's empowerment through water resource management. Women's participation is a key issue in participatory water management. So, more emphasis on women's direct involvement in water management should be placed in order to bring about gender equality through water-related projects. Gender equality in water management is still a debated issue in Bangladesh.

However, such studies as ours can make for further ground-level researches. Gender-responsive budgeting, as also gender auditing, in all water-related projects and institutions is very essential for further evaluation of projects and their issues of gender equity.

References

Ahmed, M. F. and H. Jahan. 2000. *Participatory Management of Low Cost Water Supply and Sanitation.* Dhaka: International Training Network (ITN) Centre, Bangladesh University of Engineering and Technology (BUET).

Asian Development Bank (ADB). 2006a. *NGO, Slum Dwellers, Surmount Water and Sanitation Problems in Dhaka.* Dhaka: ADB.

———. 2006b. *Country Water Action Bangladesh.* Dhaka: ADB.

Bangladesh University of Engineering and Technology (BUET). Bangladesh Institute of Development Studies (BIDS) and Delft Hydraulics. 2003. External Evaluation Report of Small Scale Water Resources Development (SSWRDSP), Local Government Engineering Department (LGED).

Batliwala, S. 1995. "The Meaning of Women's Empowerment," *Women's World,* 29: 23–36.

Biswas, T. K. 2004. Women's Empowerment and Demographic Change. Comilla, Bangladesh: Bangladesh Academy of Rural Development.

———. 1999. "Measuring Women's Empowerment Some Methodological Issues," *Asia Pacific Journal of Rural Development,* 9(2): 63–71.

Brandi, 2005. Empowerment of Coastal Fishing Communities Empowerment for Livelihood Security, BGD/97/017.

Chowdhury, A. N. 1990. *Let Grassroots Speak: People's Participation Self-help Groups and NGO's in Bangladesh.* Dhaka: University Press Limited.

Chowdhury, N. J. 2005. "Empowerment in Bangladesh: Some Concepts and Concerns," *Empowerment: A Journal of Women for Women,* 12: 17–34.

Duza, A. and H. A. Begum. 1993. *Emerging New Accents: A Perspective of Gender and Developement in Bangladesh.* Dhaka: Women for Women.

Gender and Water Alliance (GWA). 2005. *Resource Guide Mainstreaming Gender in Water Management.* The Netherlands: GWA.

Goswami, A. K. 1998. "Empowerment of Women in Bangladesh," *Empowerment: A Journal of Women for Women,* 5: 45–74.

Government of the People's Republic of Bangladesh. 1999. *Gender Dimension in Development: Statistics of Bangladesh.* Dhaka: Bangladesh Bureau of Statistics (BBS) and MoWCA (Ministry of Women's and Children Affairs).

Government of the People's Republic of Bangladesh. 1999. *National Water Policy*. Dhaka: Ministry of Water Resources (MoWR).

———. 1997. *National Women Development Policy*. Dhaka: MoWCA.

Hashemi, S. M., S. R. Sculer, and A. P. Riley. 1996. "Rural Credit Programs and Women's Empowerment in Bangladesh," *World Development*, 24(4): 635–53.

Kabir, M. R., and I. M. Faisal. 2005. "An Analysis of Gender–Water Nexus in Rural Bangladesh," *Journal of Developing Societies*, 21(1–2): 175–94.

LGED. 2003. *Gender and Development: Guidelines and Action for SSWRDSP*. Dhaka: LGED.

Nahar, B. S. 2002. *Gender, Water and Poverty: Experiences from Water Resources Management Projects in Bangladesh*. Paper presented at a Regional Workshop on Water and Poverty, September 22–26, Dhaka, Bangladesh.

Parpart, J. (ed.). 1989. *Women and Development in Africa: Comparative Perspectives*. Lanham, MD: University Press of America.

Parpart, J. L., P. M. Connelly and V. E. Barriteau (eds). 2000. *Theoretical Perspectives on Gender and Development*. Ottawa: International Development Research Centre (IDRC).

Rathgeber, E. M. 1995. *Women, Men and Water-Resource Management in Africa*. Nairobi: IDRC.

———. 1990. "WID, WAD, GAD: Trends in Research and Practice," *Journal of Developing Areas*, 24(4): 489–502.

Sen, G. and C. Grown. 1987. *Development, Crises and Alternative Visions*. New York, NY: Monthly Review Press.

United National Development Programme (UNDP). 1994. *United Development Reports on Human Development in Bangladesh: Empowerment of Women*. Dhaka: UNDP.

Visvanathan, N., L. Duggan and L. Nisonoff (eds). 1997. *The Women, Gender and Development*. Dhaka: University Press Limited.

Ward, C. 1992. "Participatory Development and the World Bank." Discussion Paper, The World Bank: Washington D.C.

PART III

Managing Groundwater

6

Impact of Electricity Tariff Policy on Groundwater Use

The Case of West Bengal, India

A. Mukherji, B. Das, N. Majumdar, B. R. Sharma and *P. S. Banerjee*

Indian policy discourse on the most suitable mode of agricultural electricity tariff has come full circle. Until the early 1970s, all state electricity boards (SEBs) charged their tubewell owners on the basis of metered consumption of electricity. However, as the number of tubewells increased manifold during the 1970s and the 1980s, the SEBs found the transaction costs of metering to be prohibitively high as compared to the total revenue generated from the agricultural sector. In response, during the 1970s and 1980s most states introduced flat tariffs for agricultural electricity supply (Shah et al. 2007). The initial idea was to increase the flat tariffs over time in order to keep them in line with the cost of generation and supply of electricity.

While this solution lowered the transaction costs of bill collection, it resulted in a set of still graver problems affecting both the electricity and the groundwater sectors. For one, many state governments soon started using the electricity tariff as an electoral tool of appeasement and hence the flat tariffs remained perpetually low (Dubash and Rajan 2001). This resulted in losses to the SEBs estimated at around INR 270 billion per year (World Bank 2002). Unmetered electricity supply also became a convenient garb for the SEBs to hide their inefficiencies in terms of transmission and distribution losses (Sant and Dixit 1996). Over time, however, the SEBs came to treat their agricultural consumers as a liability. As a result, quality of power in rural areas deteriorated and some states saw "de-electrification" and consequent stagnation in agricultural electricity consumption. In other states, where electricity consumption in agriculture grew over time (Gujarat, Andhra Pradesh, Punjab, Haryana and Tamil Nadu), the number of hours of electricity supply came down from 18–20 in the 1980s to as low as 6–10 in the 2000s. Rationing, that too of low-quality electricity, soon became the norm.

There were equally serious implications for the groundwater sector. Since the marginal cost of extracting groundwater was close to zero, it provided incentive for over-pumping. In many areas, this spawned active groundwater markets. In arid and semi-arid regions with hard rock aquifers, flat tariff was directly responsible for over-pumping and, given the low recharge potential of these aquifers, water tables declined sharply. This, in turn, put in jeopardy the livelihoods of millions of poor farmers dependent on groundwater irrigation (Moench 2007). In contrast, in areas of abundant rainfall and rich alluvial aquifers with adequate recharge during the monsoon season, for instance, West Bengal (Mukherji 2007a and 2007b), the flat tariff system did not induce over-exploitation of groundwater.

Low flat tariff and the resultant electricity subsidy have also been criticized from an equity perspective. It is often alleged that much of the agricultural electricity subsidy goes to the rural rich because they own a major proportion of the water extraction mechanisms (WEMs) fitted with electric pumps (Howes and Murgai 2003; World Bank 2002). However, this particular critique of flat tariff is not very well founded as it disregards the existence of informal groundwater markets. In a scenario of active groundwater markets, it is not the landholding size of the pump owners that matters; what matters more is the total command area of the tubewell including the area of operation of the water buyers, as we will see later in the chapter. Recent work has shown that informal groundwater markets are indeed an all-encompassing feature of Indian agriculture, and as much as 20 million hectare (ha) of land may be irrigated through these markets (Mukherji, 2008a). In most cases, these markets also have had beneficial impacts on water buyers (Palmer-Jones 2001; Shah 1993).

Nevertheless, in view of several criticisms of the flat tariff system, there is a growing pressure from the electricity utilities and the international donor agencies such as the World Bank and the Asian Development Bank (ADB) to revert to the metering of agricultural electricity supply. This is also articulated in the Electricity Act of 2003 which states: "No licensee shall supply electricity, after the expiry of two years from the appointed date, except through installation of a correct meter in accordance with the regulations to be made in this behalf by the Authority" (*Electricity Act 2003*, Article 55 [1]).

The World Bank and the ADB have also made increase in tariff, coupled with universal metering, a precondition for financing power sector reforms in any state. However, several states such as Haryana (Dubash and Rajan 2001) and Gujarat (Shah and Verma 2008) have resisted any attempt to meter agricultural power even at the cost of foregoing loans from the World Bank and the ADB respectively. The reason why these governments are unwilling to accept metering in the agricultural sector is the tremendous pressure from their rural votebank.

Thus, while the donor agencies and the Government of India (GoI) are pushing hard for metering, there are very few takers for universal metering. The state of West Bengal is an exception in this regard. As per a memorandum of understanding (MoU) signed between the GoI and the Government of West Bengal (GoWB) in 2000, the state government has agreed to universal metering of consumers (see http://powermin.nic.in, accessed October 27, 2013). In the agricultural sector, metering has been completed in 70 percent of the cases, and consumers in few districts such as North 24 Parganas, Nadia and Murshidabad have started receiving bills according to meter reading. It is envisaged that by March 2009, the goal of universal metering will be achieved.[1] West Bengal also differs from some other major Indian states in terms of both groundwater and electricity use (Table 6.1).

Table 6.1
Comparison of groundwater and electricity scenarios in West Bengal and other states

Indicators (Year)	*West Bengal*	*Punjab*	*Haryana*	*Gujarat*	*Tamil Nadu*	*Andhra Pradesh*	*Source*
Level of development of groundwater (%) in 2004	42	145	109	76	85	45	CGWB 2006
Number of over-exploited blocks as percentage of the total number of blocks in 2004	0 (0)	103 (74.6)	55 (50.9)	31 (16.8)	142 (37.0)	219 (19.8)	CGWB 2006
Normal average annual rainfall (mm)	2074	780	615	1243	995	561–1113	CGWB 2006
Nature of aquifer	Alluvial	Alluvial	Alluvial	Alluvial & hard rock	Hard rock	Hard rock	CGWB 2006
Percentage of electric tubewells to total tubewells in 2001	8.2	73.3	63.1	54.5	82.5	93.5	GoI 2003

Agricultural electricity consumption (Mkwh) in 2000–01	1360	8200	5171	14507	9066	11222	Mukherjee 2008
Percentage of the share of agriculture out of total electricity consumption in 2001–02	6.1	35.5	47.2	45.9	28	40.5	Planning Commission 2002
Flat tariff (INR/HP/year) in 2007	1760–2160	Free	420	850	Free	Free	Authors
Electricity subsidy as percentage of fiscal deficit in 2000–01	0.8	38	78	56	42	54	Briscoe 2005
Percentage of households reporting hiring irrigation services from others in 1997–98	67.2	19.3	38.5	NA	24.6	33.8	NSSO 1999

Source: As mentioned in the last column of the table.

Note: 1. NA = Not available.

Research Questions and Methodology

On the one hand, metering of agricultural electricity supply has been recommended on the grounds of efficiency (both financial and technical), equity and sustainability of the electricity utilities and groundwater use. On the other hand, it has been discouraged on the grounds of high transaction costs and its possible negative impact on groundwater markets. In this chapter, we have, therefore, attempted to answer the following research questions:

(*a*) How is the GoWB proposing to minimize the transaction costs of metering?
(*b*) How do the consumers perceive metering?
(*c*) Who would gain and who would lose under the new metered tariff regime?
(*d*) What would be the probable impact of metering on the functioning of groundwater markets?
(*e*) What would be the probable impact on groundwater extraction and use?

To answer these questions, a number of methods were adopted. To understand the current dynamics of metering, the officials of the West Bengal State Electricity Distribution Corporation Limited (WBSEDCL) were interviewed. To understand the farmers' perceptions about metering, a primary questionnaire-based survey was administered to 155 respondents in five districts of West Bengal. This is one of the two surveys undertaken as part of the Groundwater Governance in Asia (GGA) project and will be henceforth referred to as the 1st GGA 2008 survey. For defining the losers and gainers under the regime of metering, data from two additional surveys were used. The first is the 2004 survey data collected by the first author of this chapter and will be referred to as Mukherji 2004 survey. This dataset contains observations from 137 electric-pump-owning respondents spread across six districts of the state. The second data set is a qualitative survey carried out in 2008 in 17 villages spread across three districts of West Bengal. This will be referred to as the 2nd GGA 2008 survey. The specific purpose of this survey was to see how water prices and other terms and conditions of exchange in groundwater markets have changed in response to metering. Table 6.2 gives the details of all three datasets.

Metering in West Bengal: The High-tech Way

The WBSEDCL introduced high-tech meters based on GSM cellular modules in the rural areas. These meters can be remotely read from a distance of 100 feet or more, and meter readings are transferred to the regional and central commercial offices in real time. The redesigned meters are tamper-resistant, and any attempt to bypass or tamper with them is reported instantly to the central distribution office (see Figure 6.1). Power theft and tampering with meters has been also declared a serious offence under the Indian Electricity (West Bengal Amendment) Act of 2001, whereby offenders can be imprisoned for up to five years or fined up to an amount of INR 50,000, in addition to being subjected to several other punitive actions. The law was put to effect in July 2002 and from then until July 2003, 2,000 raids and 73 arrests were made (EEFI 2002).

These meters are also Time of the Day (TOD) meters. TOD is a demand-side management tool, whereby a certain section of consumers are discouraged to utilize energy during peak hours when there is a huge demand from other sectors. For this purpose, the meters have been devised with three sectors being activated rotationally in three distinctly different time slabs for agricultural pump sets. These time slots are from 0600 hours to 1700 hours (Normal "N" tariff at the rate of INR 1.37/KwH); 1700 hours to 2300 hours (Peak "P" tariff at the rate of INR 4.75/KwH) and 2300 hours to 0600 hours (Off-peak "O" at the rate of INR 0.75/KwH). The cost of the meters is to be recovered from the consumers in eight equal instalments anytime within a period of 24 months from the date of installation of the meter.

WBSEDCL has outsourced meter reading to the manufacturers of TOD meters on a contract basis for an initial period of two years. About 300 members from 100 Self Help Groups (SHGs) with members mostly from backward castes are now being trained

Table 6.2
Details of data used

	Mukherji 2004 Survey	*1st GGA 2008 Survey*	*2nd GGA 2008 Survey*
Nature of data	*Quantitative*	*Quantitative*	*Qualitative*
Time of survey	Aug.–Dec. 2004	Jan.–Feb. 2008	Mar.–Apr. 2008
No. of districts covered	6	5	3
No. of villages covered	22	15	17
No. of respondents	137	155	143
No. of electric pump owners	137	108	71
No. of submersible pumps	65	86	–
No. of centrifugal pumps	72	22	–
No. of pump owners who do not sell water	7	8	–
No. of water sellers	130	101	71
No. of water buyers	0	47	72
Names of the districts covered	Birbhum, Bardhaman, Hugli, Murshidabad, Nadia, North 24 Parganas	Bankura, Bardhaman, Hugli, Nadia, North 24 Parganas	Murshidabad, Nadia, North 24 Parganas
Data used for	Defining the losers and gainers under metered tariff system among the electric pump owners	Understanding perceptions of pump owners, water sellers and water buyers regarding metering	Defining the losers and gainers among water buyers

Source: Primary survey by the authors.

Figure 6.1
A schematic diagram of a generic IT Power Distribution System used in West Bengal

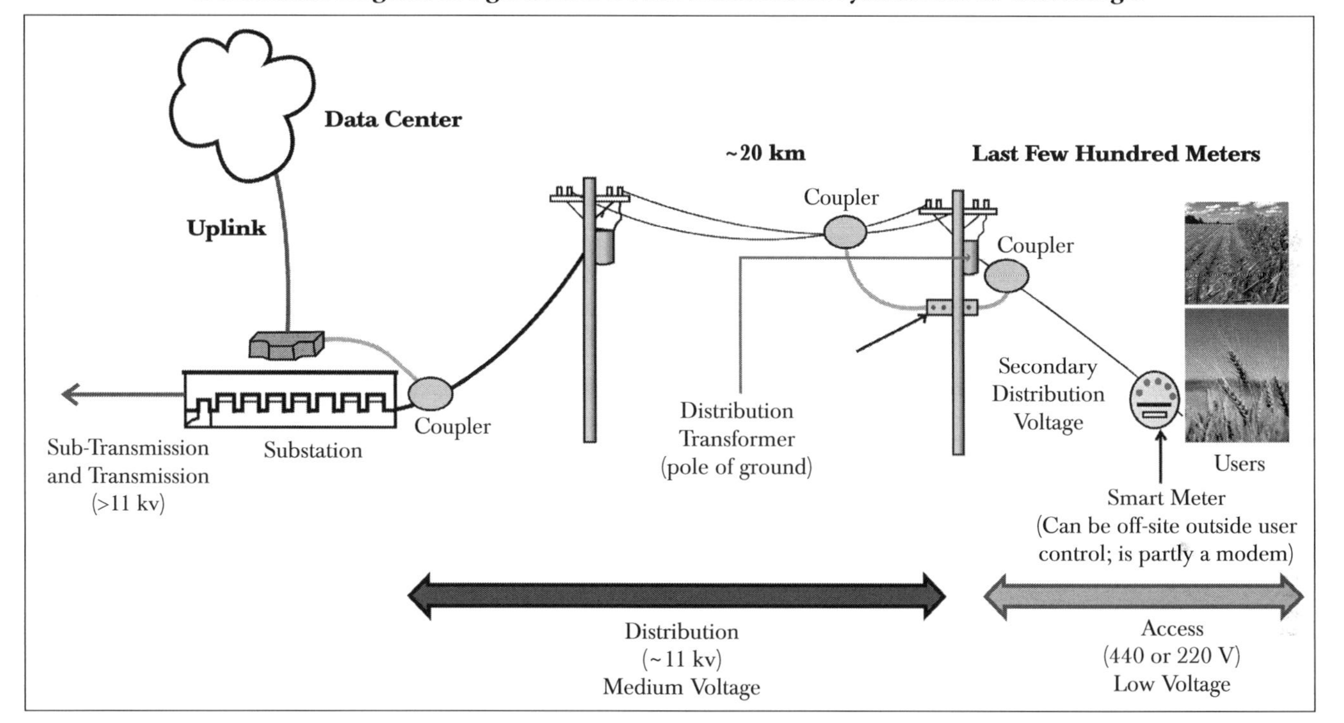

Source: Adapted from Tongia 2004.

by the WBSEDCL for meter reading, billing, petty repair, collection of revenues, mobilization of prospective consumers, etc. (*Vidyut Baarta* 2007).

Perception of Pump Owners and Water Buyers

In West Bengal, individual landholding size of pump owners is small and, therefore, self-irrigation alone did not justify the high electricity bills under flat tariff system. Hence, under a high flat tariff regime as it existed before, the pump owners became pro-active water sellers. Over time, as the flat tariff increased, the water charges did not increase proportionately. This meant that their profit margins from water selling declined over time. Since the water buyers too realized that water sellers were under a compulsion to sell water to them, their bargaining power increased considerably and they were able to get concessions from the water sellers in terms of lower water rates, deferred payment facilities, etc. (Mukherji 2007a, 2008b). It is, then, not surprising that most pump owners did not like the flat tariff system and, under the umbrella of All Bengal Electricity Consumers Association (ABECA), successfully lobbied with the government for installation of meters. Given this background, 155 respondents from five districts in West Bengal were asked about their preference regarding the mode of tariff. The response of the pump owners and water buyers was quite predictable. Of the pump owners, 67 percent preferred metered tariff system, while only 25.5 percent of the water buyers preferred the same. Table 6.3 shows the preferences of the respondents.

Who Gains and Who Loses?

In this section, we have used two sets of data (Mukherji 2004 survey and 2nd GGA 2008 survey) to delineate the losers and gainers among pump owners and water buyers under the new metering system. A pump owner can be considered a loser under the new meter tariff system if s/he has to pay a higher electricity bill than what s/he was paying under flat tariff system for the same number

Table 6.3
Preferences of respondents regarding the mode of electricity tariff estimation

Ownership Status	*Sample Size*	*Do You Prefer Metered Tariff Over Flat Tariff?*		
		Yes	*No*	*Can't Say*
Pump owner	108	73 (66.9)	27 (25.7)	8 (7.4)
"Pure" water buyer	47	12 (25.5)	24 (51.1)	11 (23.4)

Source: 1st GGA 2008 survey.
Note: Figures in parentheses are percentage to the total.

of hours of operation. A water buyer is considered a loser, if s/he has to pay higher water charges under the new meter tariff system than under the old flat tariff system for using the same amount of water, or s/he receives a poorer quality of service or more adverse terms of contract than what s/he received under the old system. The electricity utility is deemed to be a loser if the amount of revenue generated from the same number of agricultural consumers under the metered tariff is lower than the revenue collected previously under the flat tariff regime.

The Pump Owners

On an average, a 5 HP pump consumes 3.73 units of electricity per hour of operation (at the rate of 0.746 KWh/HP). Given the different tariff rates, the average electricity bill works out to INR 5.54 per hour. To this, meter rent at the rate of INR 22 per month per connection is added. Based on the number of hours of operation of a pump in a day and the type of crop grown and assuming that the pump owners would operate their pumps for the same number of hours under the metered tariff system as they did under the flat tariff system, metered bills for our sample tubewells were calculated.

Under the current meter tariff rates, it was found that out of 65 submersible pump owners, 41 (or 63.1 percent) would pay a lower electricity bill under the metered tariff system than they were paying under the flat tariff system. Out of the centrifugal pump owners, 73.6 percent would be "winners" as against 26.4 percent "losers." Figures 6.2a and 6.2b show the gainers and losers for submersible and centrifugal pumps respectively.

Figure 6.2a
Predicted electricity bill at current metered tariff rates versus electricity bill at flat tariff rates for submersible pump owners in West Bengal (N = 65)

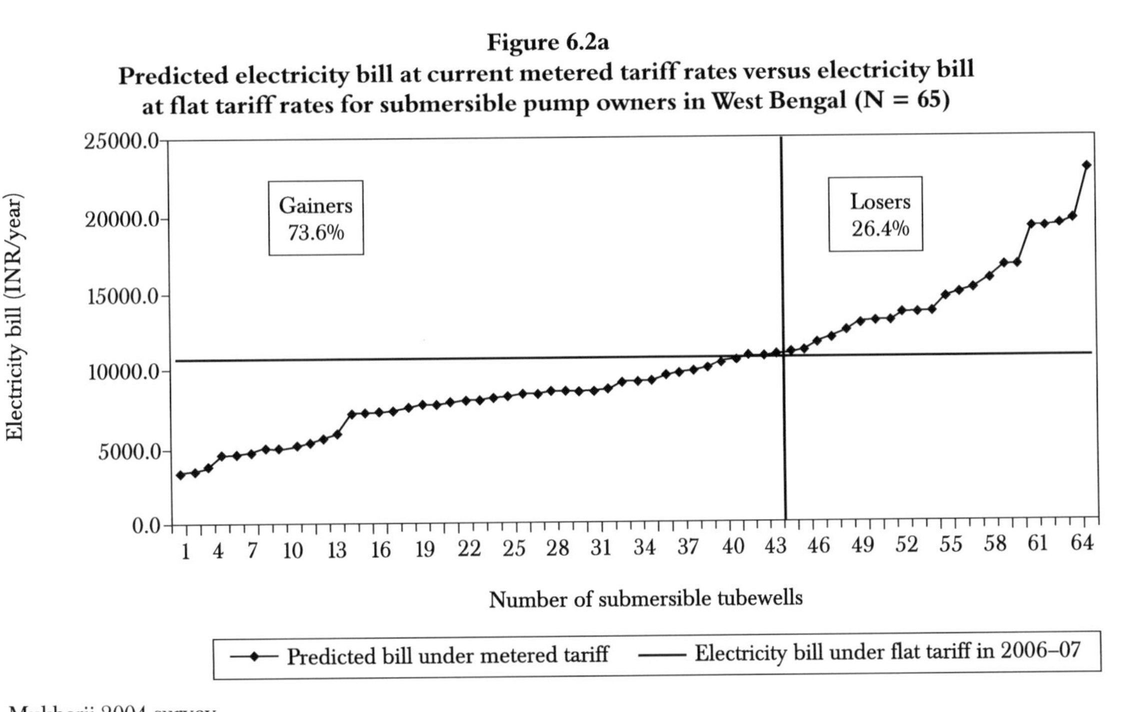

Source: Mukherji 2004 survey.

Figure 6.2b

Predicted electricity bill at current metered tariff rates versus electricity bill at flat tariff rates for centrifugal pump owners in West Bengal (N = 72)

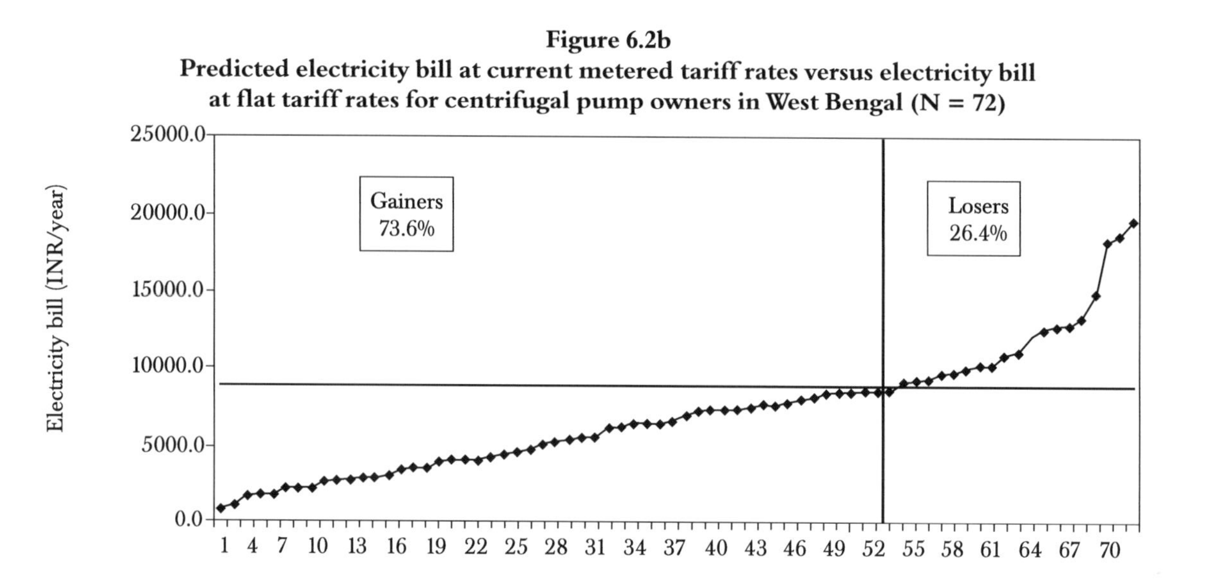

Source: Mukherji 2004 survey.

At the existing tariff rates, the average metered tariff works out to be INR 1.48/KWh. However, the cost of power supply for the West Bengal State Electricity Board (WBSEB) was INR 3.77/unit in 2001–02 (Planning Commission 2002). It is to be expected that the West Bengal State Electricity Regulatory Commission (WBSERC) will keep revising its tariff rates upwards and try to bridge the gap between cost of production and supply of electricity. Quite predictably, as the tariff rate goes up, the number of losers according to our definition will increase.

Water Buyers

Water buyers would lose out under the new metered tariff regimes, if: (*a*) the price at which they buy water goes up; (*b*) water sellers show unwillingness to sell water; and (*c*) if other terms and conditions of water sale become unattractive for the buyers. The 2nd GGA 2008 survey was specifically aimed at capturing village-level changes in the terms and conditions of water sales after the metering of agricultural tubewells was put in place.

It was found that in all the villages without any exception, water rates for all crops have increased by 30–50 percent after the introduction of the metered tariff rates. In West Bengal, usually, three modes of payment are found. These are: (*a*) crop- and season-wise cash contract (INR/*bigha*[2]/crop); contracts for *aman* and *boro* paddy[3] are of this kind; (*b*) hourly rate (INR/hour), which is common for all other crops; and (*c*) one-time crop- and area-specific contracts (INR/*bigha*/irrigation) which is usually found in case of crops with low water requirement such as mustard, wheat and sesame. Table 6.4 shows the increase in water rates for different types of crops after the introduction of metered tariff rates.

This increase in water price is not in anticipation of higher electricity bill, as we have shown that under the current tariff system, 63 percent of submersible owners and 76 percent of the centrifugal pump owners would have to pay a lower electricity bill than they did before. According to the water buyers, the reason for increasing water charges was the changed incentive structure for the pump owners. Unlike the case under the system of high flat tariffs, under

Table 6.4
Change in water rates after introduction of metering

Crop	*Unit*	*Water Rates Under Flat Tariff System in 2006–07*			*Water Rates Under Metered Tariff System in 2007–08*			*% Increase*
		Min.	*Max.*	*Av.*	*Min.*	*Max.*	*Av.*	
Aman paddy	INR/bigha	350	800	500	500	1000	660	32
Boro paddy	INR/bigha	600	1200	850	800	1500	1100	29.4
Any crop	INR/hour	15	40	25.8	25	50	37	43.4
Wheat, Mustard, Sesame	INR/irrigation/bigha	50	200	97.6	100	300	150	53.7

Source: 2nd GGA 2008 survey.

metered tariff system, they are no longer under a compulsion to sell, and as a consequence the bargaining power of water buyers has declined. A water buyer in a village in Murshidabad district articulated this issue of changed incentive structure well:

> Till last year, my water seller would come to my house before the boro season just to make sure that I would buy water from him for the season. I usually paid at the end of the season. This year, he increased the water charges from ₹800/bigha to ₹1200/bigha. I objected. He told me that I can buy water from him if I wanted to; otherwise I can go somewhere else because now that he has a meter, he will not bother much about selling water. He also asked for an advance of ₹300 saying he now needs to pay his electricity bill every month (English translation of an interview in Bangla with a water buyer in a village in Murshidabad on April 4, 2008, 2nd GGA 2008 survey).

Groundwater Markets

In 1999, there were 6.1 million farming households in West Bengal, of which only 1.1 million households owned water extraction mechanisms, while another 3.1 million households reported hiring irrigation services from their neighbours (NSSO 1999). There is evidence to show that recent expansion in groundwater markets has been a direct result of the steep rise in flat tariff rates. Earlier, i.e., in the early 1990s when the flat tariff rates were low, pump owners were more interested in leasing in land from the prospective water buyers than selling water to them (Webster 1999). However, Mukherji (2007a, 2007b and 2008b) shows that in recent years, high flat tariff rates provided a positive incentive to the pump owners to sell water and, in the process, recover their electricity bills, as also earn additional profits.

What would happen to the size and intensity of groundwater market transactions as result of the metering of electricity supply? Given that the incentive to sell water to others, as was present under the flat tariff system, is missing under the metered tariff system, those pump owners who were not overly driven by profit motive and yet were under compulsion to sell water just to recover their electricity bills, would possibly exit from the market. Under this scenario, water markets would shrink in size. From our 2nd GGA 2008 survey, we found that in four out of 17 villages, area under boro paddy cultivation had declined in 2007–08 in response to the hike in water rates for boro paddy. This happened in spite of the fact that paddy prices were at their historic high during that year. In these villages, the depth of water market transactions would have certainly gone down. Similarly, we also found that in 10 out of 17 villages, pump owners had shown greater interest in leasing in land from their erstwhile water buyers instead of selling water to them. In these villages too, both breadth and depth of water markets are likely to have gone down.

On the other hand, pump owners who were motivated by the profit they made from selling water, would continue to do so and get a higher profit margin due to lower cost of pumping and high water rates under the metered tariff system. If pump owners are driven predominantly by the profit motive, water markets might even expand in the long run or at least remain constant. But this is an empirical question that needs to be studied carefully.

Groundwater Use

One of the most important assumptions behind marginal cost pricing of water or electricity is that it would reflect the scarcity value of water and, therefore, increase water use efficiency. Evidence for this, however, is at best sketchy and even contradictory (Venot and Molle 2008; Kishore and Verma 2004; World Bank 2002). Thus, whether or not metering of pumps in West Bengal would lead to reduction in pumping or increase in water use efficiency is an empirical question that can be answered only when we have comparative data in the future. However, data from our 2nd GGA 2008 survey shows that there has been no change in gross irrigated area in any of the villages, though in four out of 17 villages, area under water-intensive summer boro paddy has declined in response. In these villages, it is likely that groundwater extraction would have reduced, but not so in other villages. However, we did find evidence that pump owners are trying to minimize seepage losses by using rubber pipes (in 10 out of 17 villages, rubber pipes were used for the first time after metering), by maintaining their unlined channels better and, in some isolated cases, by constructing underground channels. Water use efficiency, therefore, might go up, but whether or not it will lead to conservation of groundwater is a tricky question. There is also no evidence to show that quality of electricity supply, which was relatively satisfactory in the past, has improved in response to metering.

Electricity Utilities

The Memorandum of Understanding (MoU) signed between the GoWB and GoI in 2001 states that "the reform measures are being undertaken with the objective of achieving break-even in the SEB by March, 2003 and getting positive returns thereafter" (http://powermin.nic.in). According to a statement made in the West Bengal Assembly in 2006, reform measures have led to a turnaround in the financial performance of the SEB, i.e., from a loss of INR 5.2 billion in 2001–02 to a commercial profit of INR 0.81 billion in 2005–06 (http://siteresources.worldbank.org). It is to be noted that from 2001 to 2006, the flat tariff for agriculture increased from INR 3,350/year to INR 8,950/year for centrifugal pumps and INR 5,031/year to INR 10,930/year for submersible pumps. This increase contributed in part to higher revenues for the SEB. At the current metered rates, and assuming same hours of pumping, it is likely that WBSEDCL will lose out in terms of revenues in the short term. But this will change once tariff rates are increased.

Conclusion and Policy Implications

While universal metering is often thought to be a panacea of all ills in the electricity and groundwater sectors, high transaction costs have often thwarted such an expectation. The initiative of the GoWB in this regard is quite innovative. In terms of the design of the program, the GoWB has adopted a hi-tech. approach aimed at reducing the transaction costs of metering. The introduction of GSM-based electronic and remotely read meters with tamper-proof properties takes care of many of the conventional shortcomings of metering in rural areas.

Metering is often advocated on the grounds that it would be beneficial to both farmers and state electricity utilities. Our study found that metering has indeed been beneficial to the pump owners in two ways: first, they have been able to charge higher water charges for same amount of water sold and second, they have to pay lower electricity bill than before. The electric pump owners number just above 1,00,000 and hence constitute less than 2 percent of the agricultural households in the state. They also happen to be larger and wealthier farmers (Mukherji 2007a).

Water buyers, on the other hand, have lost out under the new metered tariff system in several ways. First, they now have to pay a higher price for buying water. Second, their bargaining power vis-à-vis the water sellers has declined considerably, and as a result they are now being forced to buy water at disadvantageous terms and conditions. This unwillingness of the pump owners to sell water is manifest in their eagerness to lease in land from the erstwhile water buyers. If this occurs, it will make the current water buyers increasingly dependent on the market for procuring food grains for self-consumption or push them out of farming. Water buyers constitute 50 percent of the rural farming households (NSSO 1999) and often belong to poor and marginal sections of the society.

Under the existing electricity tariff system, even the state electricity utilities are likely to lose out in terms of revenues, but this will change as soon as the tariff rates are revised upwards. Another justification for metering is that it will lead to better energy auditing. However, better energy auditing can be also done at transformer level, instead of individual meters. That the largest section of the rural community, namely, the water buyers, has been negatively affected by metering also calls for questioning the assertion that metering will improve the lives of India's farmers (World Bank 2002).

Marginal cost pricing through metering might lead to improved water use efficiency and the results would be positive. However, whether or not it will lead to water savings is a debatable issue. More debatable is whether or not conservation of groundwater should be the prime policy objective in a state that is flushed with groundwater and steeped in poverty and where groundwater may be used for poverty alleviation (Kahnert and Levine 1993).

Metering and, therefore, proper auditing and accounting of energy supply are also thought to ultimately improve the quality of electricity. However, as mentioned earlier, farmers in West Bengal receive relatively high-quality electricity supply and, during our survey, we did not find any evidence for further improvement in quality consequent upon metering.

Given that the GoWB has already invested millions of rupees in metering and that the lending agencies also insist on it as do India's national policies, it is unlikely that metering will be revoked. Under such a scenario, what are the policy options that might soften the blow to the poorer water buyers?

The GoWB needs to take steps to accelerate the pace of electrification of tubewells in the state. This will enhance competition in the water markets and, in response, water prices might decline. On the positive side, metering of electricity would encourage many small farmers – who earlier might have been reluctant to invest in tubewells fearing that they would not be able to recover the high flat tariffs through selling water – to invest in tubewells. Under the metering system, they would have to pay for only as much as they consume. However, as per the current government policies, getting a new electricity connection for a tubewell is a cumbersome process (Mukherji 2006). Besides, with phasing out of all capital subsidies from the late 1990s, construction of electric tubewells has become a costly affair requiring anything from INR 50,000 to INR 150,000 per tubewell. The GoWB should relax the stringent regulations vis-á-vis groundwater permits, as also provide one-time capital subsidy to small and marginal farmers for the construction of tubewells. This will reduce their dependence on water markets for irrigation.

The Panchayats (village-level governing bodies) can play an important role in regulating water prices in the market, by setting the maximum price at which a pump owner can sell water in the village. In some villages in West Bengal, Panchayats are already playing this regulatory role (see Mukherji 2007c; Rawal 2002), but this regulatory mechanism needs to be carefully replicated on a larger scale, keeping in mind issues of corruption and capture of such regulatory bodies by the local elite.

≈

Acknowledgements

The authors are grateful to the Challenge Program on Water and Food (http://www.waterandfood.org, accessed October 27, 2013) for funding the Groundwater Governance in Asia Project (http://www.waterforfood.org/gga/, accessed October 27, 2013). This chapter is the output of the project. An extended version of this chapter was published in *Energy Policy* in 2009 (Mukherji et al. 2009).

Notes

1. Personal communication with an official of West Bengal State Electricity Distribution Company Limited (WBSEDCL).
2. *Bigha* is a local unit of land and roughly 7.47 *bighas* is equivalent to a hectare.
3. *Aman* paddy is a rainy season paddy. It is sown in July and harvested in November. Boro paddy is winter paddy, sown in December/January and harvested in April/May.

References

Briscoe, J. 2005. *India's Water Economy: Bracing for a Turbulent Future.* Washington DC: World Bank.

Central Ground Water Board (CGWB). 2006. *Dynamic Groundwater Resources of India (as on March 2004).* Faridabad: CGWB, Ministry of Water Resources, Government of India (GoI).

Dubash, N. K. and S. C. Rajan. 2001. "Power Politics: Process of Power Sector Reform in India," *Economic and Political Weekly*, 36(35): 3367–90.

Electricity Employees Federation of India (EEFI). 2002. "The Impact of the Indian Electricity (West Bengal Amendment) Act 2001," *Voice of the Electricity Workers*, 3(3). Also available online at http://www.eefi.org/0702/070224.htm (accessed May 21, 2008).

Government of India (GoI). 2003. *All India Report on Agricultural Census, 2001–02.* New Delhi: Ministry of Agriculture, GoI.

Howes S. and R. Murgai. 2003. "Karnataka: Incidence of Agricultural Power Subsidies: An Estimate," *Economic and Political Weekly*, 38(16): 1533–35.

Kahnert, F. and G. Levine. 1993. *Groundwater Irrigation and the Rural Poor: Options for Development in the Gangetic Basin.* Washington DC: World Bank.

Kishore, A. and S. Verma. 2004. "What Determines Pumping Behaviour of Tubewell Owners: Marginal Cost or Opportunity Cost?" Water Policy

Research Highlight 6, International Water Management Institute (IWMI) – Tata Water Policy Program, Anand, Gujarat, India.

Moench, M. 2007. "When the Wells Run Dry but Livelihood Continues: Adaptive Responses to Groundwater Depletion and Strategies for Mitigating the Associated Impacts," in M. Giordano and K. G. Villholth (eds), *The Agricultural Groundwater Revolution: Opportunities and Threats to Development*, pp. 173–92. Comprehensive Assessment of Water Management in Agriculture Series No. 3. UK: CABI Publishers.

Mukherjee, S. 2008. "Decomposition Analysis of Electricity Consumption: A State-wise Assessment," *Economic and Political Weekly*, 43(3): 57–64.

Mukherji, A. 2008a. "Spatio-temporal Analysis of Markets for Groundwater Irrigation Services in India, 1976-77 to 1997–98," *Hydrogeology Journal*, 16(6): 1077–87.

———. 2008b. "The Paradox of Groundwater Scarcity amidst Plenty and Its Implications for Food Security and Poverty Alleviation in West Bengal, India: What can be Done to Ameliorate the Crisis?" Paper presented at 9th Annual Global Development Network Conference, January 29–31, Brisbane, Australia.

———. 2007a. "The Energy-Irrigation Nexus and Its Implications for Groundwater Markets in Eastern Indo-Gangetic Basin: Evidence from West Bengal, India," *Energy Policy*, 35(12): 6413–30.

———. 2007b. "Political Economy of Groundwater Markets in West Bengal: Evolution, Extent and Impacts." PhD thesis, University of Cambridge, Cambridge, UK.

———. 2007c. "Equity Implications of Alternate Institutional Arrangements in Groundwater Sharing: Evidence from West Bengal, India," *Economic and Political Weekly*, 42(26): 2543–51.

———. 2006. "Political Ecology of Groundwater: The Contrasting Case of Water Abundant West Bengal and Water Scarce Gujarat, India," *Hydrogeology Journal*, 14(3): 392–406.

———. 2004. "Groundwater Market in Ganga-Meghna-Brahmaputra Basin: A Review of Theory and Evidence," *Economic and Political Weekly*, 30(31): 3514–20.

Mukherji, A., B. Das, N. Majumdar, N. C. Nayak, R. R. Sethi and B. R. Sharma. 2009. "Metering of Agricultural Power Supply in West Bengal, India: Who Gains and Who Loses?," *Energy Policy*, 37(12): 5530–39.

National Sample Survey Organisation (NSSO). 1999. *54th Round: Cultivation Practices in India, January 1998–June 1998*. New Delhi: Department of Statistics and Programme Implementation, GoI.

Palmer-Jones, R.W. 2001. "Irrigation Service Markets in Bangladesh: Private Provision of Local Public Goods and Community Regulation." Paper presented at Symposium on Managing Common Resources: What is the Solution?, September 10–11, Lund University, Sweden.

Planning Commission. 2002. *Annual Report (2001–02) on the Working of State Electricity Boards and Electricity Departments.* New Delhi: Planning Commission (Power and Energy Division), GoI.

Rawal, V. 2002. "Non-Market Interventions in Water-sharing: Case Studies from West Bengal, India," *Journal of Agrarian Change*, 2(4): 545–69.

Sant, G and S. Dixit. 1996. "Beneficiaries of IPS Subsidy and Impact of Tariff Hike," *Economic and Political Weekly*, 31(51): 3315–21.

Shah, T. 1993. *Water Markets and Irrigation Development: Political Economy and Practical Policy.* Bombay: Oxford University Press.

Shah, T. and S. Verma. 2008. "Co-management of Electricity and Groundwater: An Assessment of Gujarat's *Jyotirgram* Scheme," *Economic and Political Weekly*, 43(7): 59–66.

Shah, T., C. Scott, A. Kishore and A. Sharma. 2007. "Energy-Irrigation Nexus in South Asia: Improving Groundwater Conservation and Power Sector Viability," in M. Giordano and K. G. Villholth (eds), *The Agricultural Groundwater Revolution: Opportunities and Threats to Development*, pp. 211–43. Comprehensive Assessment of Water Management in Agriculture Series No. 3. CABI Publishers, UK.

Tongia, R. 2004. "What IT can and cannot Do for the Power Sector and Distribution in India: Link to Reforms, Incentives and Management." Working Paper 19, Program on Energy and Sustainable Development, Stanford University, Stanford, CA.

Venot, J. P. and F. Molle. 2008. "Groundwater Depletion in the Jordan Highlands: Can Pricing Policies Regulate Irrigation Water Use?," *Water Resources Management*, 22(12): 1925–41.

Vidyut Baarta, official journal of the West Bengal State Electricity Distribution Corporation Limited (WBSEDCL) and West Bengal State Electricity Transmission Company Limited (WBSETCL), February and June issues 2007.

Webster, N. 1999. "Institutions, Actors and Strategies in West Bengal's Rural Development: A Study on Irrigation," in B. Rogaly, B. Harriss-White and S. Bose (eds), *Sonar Bangla? Agricultural Growth and Agrarian Change in West Bengal and Bangladesh*, pp. 329–56. New Delhi: Sage.

World Bank. 2002. *Improving the Lives of India's Farmers: How Power Sector Reforms will Help*? Washington DC: World Bank.

Websites

http://powermin.nic.in/acts_notification/electricity_act2003.htm (accessed May 21, 2008) for information on *Electricity Act 2003.*

http://powermin.nic.in (accessed May 21, 2008) for information on the Electricity Act, 2003, and the MoU signed between Government of West Bengal (GoWB) and GoI.

http://siteresources.worldbank.org (accessed June 5, 2008) for information on the performance of West Bengal State Electricity Board (WBSEB) as presented in the West Bengal State Legislative Assembly in 2006.

7

Are Wells a Potential Threat to Farmers' Well-being?

The Case of Deteriorating Groundwater Irrigation in Tamil Nadu, India

S. Janakarajan and *M. Moench*

This chapter deals with the degradation of groundwater resource base due to over-extraction and pollution, in turn leading to increasing rural poverty, social inequity and conflict in several parts of India, particularly Tamil Nadu. Groundwater is a crucial productive resource not only in Tamil Nadu, but India as a whole. For rural agricultural population, it has almost replaced land as a determinant of socio-economic status. Increasing access to groundwater has undermined the maintenance of tank irrigation systems and other sources of surface water and, in the process, the control over water has shifted from rural communities to individuals. While access to groundwater has never been fully equitable due to the nature of resource itself and groundwater–land nexus, land-ownership, wealth and other factors, inequity in access to water is also growing. Patterns of inequity are socially embedded and exacerbated by factors, such as inheritance patterns, access to resources and competitive deepening of wells, etc. In many cases, ownership of individual wells is now divided among many people. This often results in differential access for dominant owners and those who are less capable of exercising their partial ownership rights and, in turn, leads to conflicts. Competition and conflict are also increasing in the wake of pollution and substantial decline in water level. Falling water level is leading to competitive deepening of wells and, in many areas, large financial losses, as existing wells become dry or new, unproductive wells are drilled. In many areas, shallow dug wells, too, have dried up and farmers are now drilling multiple bores alongside or within existing dug wells. Decline in water level is also leading to a decline in sources of surface water, such as the traditional "spring" channels used to divert the sub-surface flows.

Decline in water level and pollution is affecting the availability and reliability of water supply for irrigation and other uses. Farmers have responded to this water scarcity by adopting efficient water use technologies. Nonetheless, water scarcity is reducing yields and directly impacting agricultural incomes. Indirect impacts are also major. Informal water markets, for example, initially emerged as farmers with access to water surplus supply sold water to neighboring farmers who either lacked financial resources to dig their own wells or had insufficient water in the wells they owned. Now, such informal water markets, too, are declining, as farmers reserve all the available sources of water supply they own for their own use. Furthermore, even where water markets continue to exist, their operation is often highly inequitable, since they function as part of interlocked land and labor markets where purchasers are dependent on the goodwill of water sellers. As water becomes increasingly scarce, relations of dependency intensify putting purchasers in an ever-weaker bargaining position.

What implications does all this have for policy-making in water sector? Increasing poverty due to degradation of groundwater resource base implies that government policies supporting further groundwater development in areas suffering from overdraft must be reversed. Policies, such as the supply of highly subsidized power, are particularly problematic. In addition to encouraging indiscriminate and wholesale pumping of groundwater, such policies have their benefits largely enjoyed by wealthy sections of the rural population. Overall, policies that support more equitable access to – and sustainable use of – available groundwater resources are essential. Furthermore, in areas where inequity is high and current groundwater use patterns are unsustainable, policies supporting the efforts of marginal populations to opt out of agriculture and switch over to other livelihood activities may be required. Inherent inequities in power relations within rural communities imply that "simple" legal or other reforms to directly address groundwater overdraft and pollution are likely to be insufficient.

This chapter is organized in the following manner. The first introductory section presents an overview of the growth of groundwater irrigation in India and highlights some of the problems emerging in many regions. The following section takes up a detailed case study of the water situation in Tamil Nadu illustrating the challenges emerging in the two-thirds of India underlain by hard-rock aquifers. The first major section in the Tamil Nadu case study focuses on the characteristics of groundwater irrigation and use in the Vaigai, Noyyal and Palar basins. Then, the core issue of water level decline due to the dynamic process of competitive deepening is addressed. This is followed by a discussion of the costs of well irrigation and its relationship with the costs of surface irrigation, along with an analysis of how well irrigation is accelerating the process of social differentiation within the rural society. The final section draws conclusions at both the local and all-India levels and discusses possible policy options.

Introduction

In the decades after Independence, official statistics indicated that the number of wells and area irrigated by groundwater in India had grown and was projected to continue growing at an exponential pace until the "ultimate" irrigation potential was reached in 2007, as shown in Figures 7.1 and 7.2 (Moench 1992a, 1992b; World Bank 1998). This increase in groundwater irrigation has been a major factor contributing to the increase in yields and agricultural production at an all-India level. Yields in groundwater-irrigated areas are higher, by one-third to half, than those in areas irrigated by surface water sources (Dhawan 1995). The variability of production has also declined, in large part due to the reliability of groundwater sources (World Bank 1998).

From approximately 50 million tons in the early 1950s, India's cereal production has increased steadily to 234 million tons in 2001–02 (see Table 7.1). Per capita annual availability of foodgrains also has gone up steadily over a period of time from 141 kg in 1951

Figure 7.1
Groundwater irrigation potential

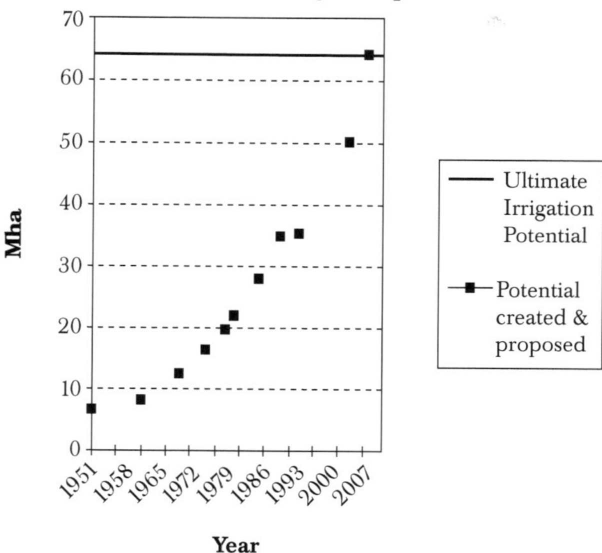

Source: GoTN 1998.

Figure 7.2
Growth of wells in India

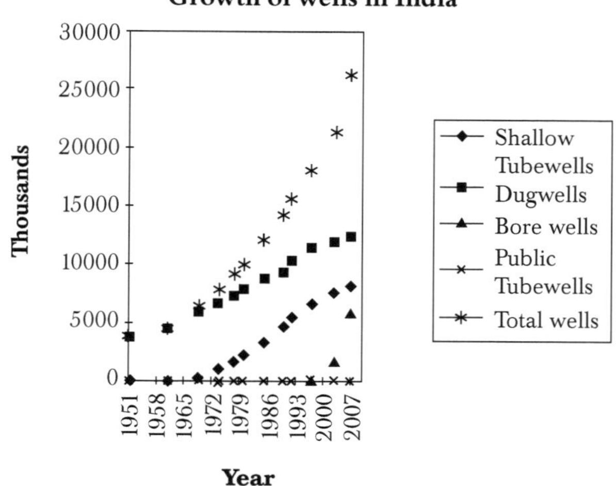

Source: GoTN 1998.

Table 7.1
Total foodgrain production in India, from 1959–60 to 2001–02

Year	*Foodgrain Production (in million tons)*
1950–51	50.8
1959–60	74.7
1960–61	79.3
1970–71	108.4
1975–76	121
1980–81	129.6
1985–86	150.4
1990–91	176.4
1991–92	168.4
1992–93	179.5
1993–94	182.1
1996–97	199.3
2001–02	234 (targeted)

Source: Center for Monitoring Indian Economy: Agriculture, November 2000.

to 200 kg in 2000. Rice and other cereals are now being exported as well. Nevertheless, food production has not resulted in food availability for all sections of society. While there is a strong association between levels of groundwater development and reduction in poverty, inequity in access to groundwater remains a major problem and sustainable future growth in food production is threatened by growing overdraft and other problems related to groundwater (Moench 2001, 2002). While India has been able to create and maintain a large buffer stock of foodgrains, a variety of concerns are emerging both at the global level and within India. According to Rosegrant and Ringler, at the global level, "the growth rate in irrigated area declined from 2.16 percent per year during 1967–82 to 1.46 percent in 1982–93. The decline was slower in developing countries, from 2.04 percent to 1.71 percent annually during the same periods" (1999). The rates of increase in yields are also declining and are projected to continue to decline in the coming decades (FAO 2000; Rosegrant and Ringler 1999). Furthermore, in some areas such as in Sri Lanka and in the rice–wheat regimes of India, Nepal, Pakistan and Bangladesh, yields have been stagnant for a number of years (Ladha et al. 2000; Amarasinghe et al. 1999). This stagnation in yields may largely be owing to emerging groundwater problems, particularly overdraft and pollution. We do not, however, believe that the relation between stagnation in yields and groundwater problems is a simple one. Instead, the impact of groundwater problems on yields and agricultural production – and more importantly, on rural livelihoods – is embedded in issues of differential access to groundwater resources that are exacerbated by unsustainable development and power relations at the village level. As one of the authors argues elsewhere, in recent decades, groundwater has played a crucial role in creating stable social conditions, conditions which are now threatened by degradation of the resource base (Moench 2002).

The situation in Tamil Nadu illustrates many of the issues that are now emerging in many hard-rock aquifers of India. Increasing number of wells is not, as Figure 7.3 demonstrates, equivalent to an increase in groundwater-irrigated area.

Figure 7.3
Well irrigation in Tamil Nadu

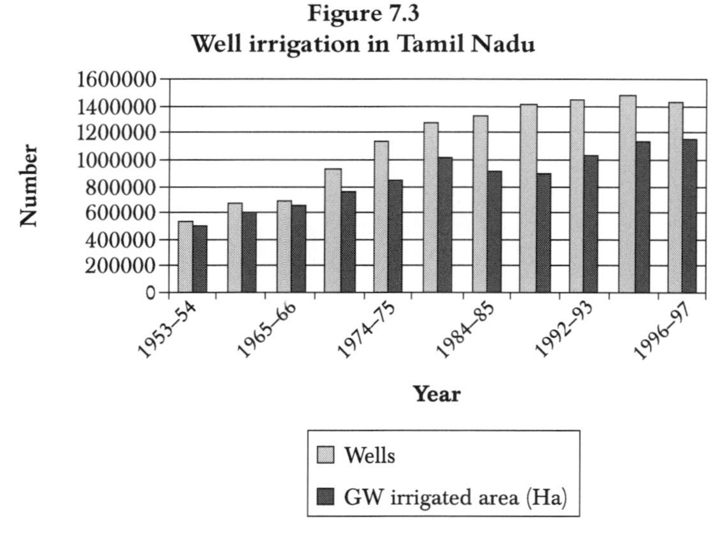

Source: GoTN 1998.

The increase in the number of wells in Tamil Nadu is following a logistic pattern, with the exponential growth rates of the 1950s through 1980s now slowing or even declining. Furthermore, although the number of wells has been increasing, groundwater-irrigated area has stagnated since the early 1980s. This pattern has emerged despite the presence of a system of extensive subsidies encouraging continued expansion of groundwater development.

What are these subsidies? The most important of them is the provision of free power to agricultural pumpsets. This cost the national exchequer approximately INR 20 billion in 1999. Other subsidies include provision of low-interest loans for deepening the existing wells or constructing new ones, for purchasing pumps and other equipments. The power subsidy has encouraged high levels of groundwater pumping and is widely held as a factor contributing to the emerging problem of groundwater overdraft (World Bank 1998; Malik 1993; Moench 1993). Well development subsidies have also had a significant impact. Despite the presence of well spacing regulations, a study undertaken in the Vaigai basin of Tamil Nadu indicates that there are now at least three wells located within the prohibited distance from every sample well selected for the survey (Janakarajan 1997a). Furthermore, while subsidy schemes have encouraged groundwater development, very little attention has been devoted to the maintenance of traditional irrigation sources, such as tanks and "spring" channels. Table 7.2 provides data on the trends in the expansion of net area irrigated by different water sources in India, and explains the fact that while well irrigation has increased manifold, the area irrigated by the conventional sources, such as tanks, is on the decline.

These water sources have played a key role not only in providing irrigation water for several centuries, but also in recharging groundwater and thereby preserving local environmental systems. Now, however, many tanks and "spring" channel sources have dried up, become clogged with silt or been encroached upon for construction or cultivation. Finally, in some other parts of the state (such as in the Palar basin), channels are being used to drain industrial effluents, despite the presence of pollution control laws. Overall, while substantial attention has been devoted to promoting groundwater irrigation, little or no attention has been paid to creating effective avenues for sustaining the groundwater resource base or for mitigating the adverse impact of groundwater depletion and pollution on rural society.

Groundwater is a crucial productive resource. Our research in Tamil Nadu indicates that access to it has almost replaced land in

Table 7.2
Trends in the expansion of Net Irrigated Area (NIA) through different sources in India, from 1950–51 to 1996–97 (area in million hectares)

	1950–51 to 1959–60		*1960–61 to 1969–70*		*1980–81 to 1989–90*		*1996–97*	
Sources	*Area*	*% of NIA*	*Area*	*% of NIA*	*Area*	*% of NIA*	*Area*	*% of NIA*
Canals	9.2	41.2	11.2	41.9	16.3	38.3	17.4	31.5
Tanks	4.2	18.6	4.5	16.6	3.0	7.0	3.3	6.1
Wells	6.6	29.8	8.7	32.6	20.8	48.7	30.8	55.9
Other sources	2.3	10.4	2.4	8.9	2.5	6.0	3.6	6.6
Total NIA	22.3	100.0	26.8	100	42.6	100	55.1	100

Source: GoI n.d(b)., in Vaidyanathan (ed.) 2001, and Centre for Monitoring Indian Economy (CMIE), September 1998.

determining one's socio-economic and political status (Janakarajan 1992, 1997a). In the past, when surface water was the only source of irrigation, the single most important productive resource was land. At that time, access to land determined one's power as well as socio-economic status in a rural society. The rapid growth of groundwater irrigation, change in cropping patterns from drought-tolerant but relatively low-yielding varieties to water sensitive but high-yielding varieties, and declining status of traditional surface water sources have resulted in the emergence of groundwater as a crucial productive resource. Therefore, in a changing agrarian context, it is the ownership of wells, along with land, that determines one's status. In Tamil Nadu, marginal and small farmers own 60 percent of the wells (Janakarajan 1997a). Ownership of wells is, however, counts for nothing unless the wells are productive and can be maintained. As a result, ownership of 60 percent of wells by small and marginal farmers does not mean greater access to groundwater resources. Declining water levels create a situation in which only those who can afford to compete in deepening their wells can have assured access to groundwater. Growing inequity in access to groundwater leads to a process of continued social differentiation which, in turn, results in deprivation, poverty and consolidation of inequitable power relations within local communities. In the sections that follow, these issues are the focus of detailed analyses based on field data collected in Tamil Nadu.

Groundwater Ownership and Access in Tamil Nadu

Groundwater access depends on a wide variety of factors, but one of the most important ones is well-ownership and the ways this factor interacts with social relations and power structures in a rural context. Rights of access to groundwater are fundamentally different from rights of access to traditional community-managed or state-managed surface irrigation systems. Under British Common Law, the basic civil law doctrine governing property ownership in most of India, groundwater rights are appurtenant to land (Singh 1991, 1990). If you own land, you can drill or dig a well and extract as much groundwater as you are able for use on overlying lands.

When land is sold, groundwater access rights pass with the land and cannot legally be separated from it. Formal legal definitions of rights, however, are often quite different from the practical "rules in use" that determine the effective access any individual may or may not have to groundwater. In Tamil Nadu, some of the most important factors affecting access to groundwater are whether wells are owned by individuals or held jointly, and ownership of wells across different categories of land-owners. These ownership factors are affected by well density, area irrigated by wells in relation to area irrigated by surface water sources, cropping patterns and yield performance.

Ownership of Wells

Sole and Joint Ownership of Wells

In Tamil Nadu, agricultural land is generally divided between heirs at the time of inheritance. Increasingly, this is also the case with wells. Because landholdings are relatively small, water is a critical resource and wells are key productive assets, ownership of wells is often split into shares at the time of inheritance. As a result, wells in Tamil Nadu are increasingly shifting from single owners to joint ownership. This is of fundamental importance for understanding the emerging groundwater problems and potential solutions because it has become a central point of conflict within communities and families. Joint ownership is increasing the rate of differentiation between the "haves" and the "have-nots." Sometimes, the results are extreme; after inheriting a share in a well individuals often deepen their own portion and effectively exclude other shareholders from access to water. These types of micro-level conflicts complicate the state's policy-making and appear to undermine the possibilities for arriving at a consensus on the sustainable use of groundwater resource base.

Incidence of Joint Well-ownership

The association of groundwater ownership with land ownership in combination with inheritance laws has encouraged sub-division and fragmentation of wells along with land. As opposed to single ownership of wells, virtually there is no macro-database documenting the nature and extent of joint well-ownership. Village-level studies conducted in various river basins in Tamil Nadu by the first author, however, indicate the widespread nature of joint ownership and highlight the dilemmas and uncertainties associated with the management of jointly owned wells.

Joint ownership of wells is common in Tamil Nadu. Data collected in a survey of 1,100 wells in 27 villages of the Vaigai river basin (in southern Tamil Nadu) indicate that on an average, about one-third of the wells were found to be jointly owned in that area (Janakarajan 1997a). Higher levels of joint ownership (47 percent of the sample) were found in another survey of 11 villages in the Palar river basin (Janakarajan 1999). Research conducted in the Noyyal and Palar river basins for the Local Water Management Project also shows a high incidence of joint well-ownership (see Tables 7.3 and 7.4). Of 7,120 wells surveyed in 51 villages covered by the meso-level survey in the Palar basin, the overall percentage of jointly owned wells was 43.6. The extent of joint ownership was not, however, uniform across villages. At the village level, joint ownership varied from 17.2 percent to 59.1 percent. Variation was found to be even higher in the Noyyal basin. Of the 14,358 wells surveyed in 41 villages, 53 percent were found to be jointly owned, and at the village level joint well-ownership varied from 31.3 percent to 87 percent. Joint ownership of wells is a complicated phenomenon with the number of shares in any individual well varying from a minimum of two to as many as 30 in the Palar and Noyyal basins (see Tables 7.3). There is some indication that jointly owned wells are more likely than individually owned wells to be in disuse. For the Palar basin as a whole, the percentage of joint wells in disuse was 30.4 percent, whereas, it was only 24.7 percent in the case of individually owned wells. This pattern was, however, not prevalent in the majority of villages in either the Palar or Noyyal basins.

Share Ownership of Wells Across Different Size Categories of Farmers

In addition to the widespread extent of shared well-ownership, data from eight sample villages in the Palar basin indicates that the

Table 7.3
Well-ownership patterns in Palar and Noyyal basins (1998–99)

Cluster Number	*Villages in Each Cluster*	*Number of Wells*	*Individually Owned Wells*	*Jointly Owned Wells*	*% of Individually Owned Wells*	*% of Jointly Owned Wells*	*Maximum Number of Shares in Wells*
PCluster 1	2	499	302	197	60.6	39.4	9
PCluster 2	21	2,803	1,779	1,024	63.5	36.5	29
PCluster 3	5	476	270	206	56.7	43.3	10
PCluster 4	8	1,666	681	985	40.9	59.1	8
PCluster 5	13	1,006	427	579	42.4	57.6	8
PCluster 6	2	670	555	115	82.8	17.2	5
All Clusters in Palar	51	7,120	4,014	3,106	56.4	43.6	
NCluster 1	4	1,819	1,250	569	68.7	31.3	10
NCluster 2	5	1,225	781	444	63.8	36.2	9
NCluster 3	2	438	57	381	13	87	15
NCluster 4	2	510	190	320	37.3	62.7	15
NCluster 5	7	4,610	2,112	2,498	45.8	54.2	9
NCluster 6	4	1670	325	1,345	19.5	80.5	5
NCluster 7	6	1,841	854	987	46.4	53.6	30

(*Table 7.3 continued*)

(*Table 7.3 continued*)

Cluster Number	*Villages in Each Cluster*	*Number of Wells*	*Individually Owned Wells*	*Jointly Owned Wells*	*% of Individually Owned Wells*	*% of Jointly Owned Wells*	*Maximum Number of Shares in Wells*
NCluster 8	6	634	335	299	52.8	47.2	11
NCluster 9	5	1,611	829	782	51.5	48.5	20
All clusters in Noyyal	41	14,358	6,733	7,625	46.9	53.1	

PCluster 1	Upper reach of the Palar basin where no tannery effluents are found.
PCluster 2	Upper reach of the Palar basin where tanneries are concentrated.
PCluster 3	Upper reach of the Palar basin where tanneries are concentrated.
PCluster 4	Middle of the Palar basin where tanneries and industries are concentrated.
PCluster 5	Middle of the Palar basin where tanneries are concentrated.
PCluster 6	Middle of the basin where tanneries are not located.
NCluster 1	Villages in Noyyal basin along the Tiruppur–Avinashi road.
NCluster 2	Villages in Noyyal basin along the Tiruppur–Perumanallur road.
NCluster 3	Villages in Noyyal basin along the Tiruppur–Uthukuli road.
NCluster 4	Villages in Noyyal basin along the Tiruppur–Kangayam road.
NCluster 5	Villages in Noyyal basin along the Tiruppur–Dharapuram road.
NCluster 6	Villages in Noyyal basin along the Tiruppur–Palladam road.
NCluster 7	Villages in Noyyal basin along the Tiruppur–Mangalam road.
NCluster 8	Villages in Noyyal basin along the Tiruppur–Orathapalayam road.
NCluster 9	Villages in Noyyal basin around Chennimalai textile units.

Source: Meso-level survey (1997–98).

size of shares is strongly associated with the extent of land owned by an individual farmer (see Table 7.4).

Data in Table 7.4 highlight the skewed distribution of well-ownership and its strong association with land-ownership. The key points to be noted in this data are:

(*a*) The average number of wells owned in each size class increases as size class increases. This implies that better access to land is associated with better access to groundwater.
(*b*) There is a negative association between the extent of land-ownership and the incidence of joint well-ownership. Big landowners tend to own wells in entirety rather than shares in wells. This indicates either that they construct their own new wells or that they consolidate their shares in wells by purchasing shares from other shareholders.
(*c*) Unlike big landowners, small land-owners frequently own relatively small shares in wells. All sample farmers owning less than 20 percent shares in wells are concentrated in landowning classes having less than six acres of land. This suggests that small land-owners are likely to be more vulnerable than others to losing access to groundwater.

In principle, shared ownership of wells should enable sections of society that are cannot afford to construct their own wells to obtain access to groundwater. Operation of shared wells is, however, often complicated by caste and other social factors. While we have not documented the details of the management of jointly owned wells for every case in the surveyed villages, our interviews suggest that the incidence of conflict in the sharing of water from jointly owned wells is widespread and that practical difficulties surrounding pumping and management of water shares are the most important sources of this conflict. The nature and consequences of conflict are rooted in the nature and operational practices associated with joint wells.

Joint wells are commonly operated by installing a single pump-set and running the motor in rotation between shareholders for a fixed number of hours. Operational costs are divided among shareholders in proportion to the number of shares they own.

Table 7.4
Shared ownership of wells in Palar basin

Land Holding (acres)	*No. of Households Reporting*	*% of Ownership in Shared Wells*							
		<0.1	*0.1–0.2*	*0.21–0.30*	*0.31–0.50*	*0.51–0.75*	*0.76–0.99*	*1 and More*	*Total Number of Wells Owned in the Size Class*
Upto 1	89	15	25	15	23	0	0	14	31.7
1.01–2	124	1	16	17	58	1	0	39	67.8
2.01–4	92	7	3	7	35	2	0	53	72.3
4.01–6	40	0	2	7	7	1	0	35	41.1
6.01–10	37	0	0	5	10	1	1	29	36.4
10–15	16	0	0	0	3	0	0	19	20.5
Over 15	8	0	0	1	0	1	0	12	1.3

Source: Main survey (1998–2000).

Lack of cooperation in sharing costs and the available water/ power supply are common problems. Unlike the disintegration of traditional tank irrigation communities (primarily due to the lack of incentives for tank management [Janakarajan 1993]), financial constraints is the most common problem in the installation and operation of jointly owned wells. In cases where shareholders do not contribute their portion of the costs, they are excluded from use of the pumpset. Many disputes also occur due to the erratic power supply which disrupts schedules for sharing the available pumping time. Village Panchayats (informal village assemblies) are often involved in resolving such disputes, but resolutions to these disputes are often not sustainable and conflicts re-emerge in the subsequent periods of scarcity.

An alternative to shared ownership and use of one pumpset on a jointly owned well is for each shareholder to install his own individual electricity- or diesel-operated pumpset. This is possible because most wells are dug structures with large diameters where the installation of multiple pumps is possible. However, this approach often leads to competition over available supply. Stored water is drained rapidly and competition intensifies when shareholders install high-powered motors in order to extract water rapidly. Disputes are particularly common when wells are shared by different castes. Such disputes are often only resolved when one shareholder buys the others out. In some instances, this is accomplished with poor farmers selling their lands along with their shares in a well.

In addition to disputes over pumping, disputes often occur over the need to deepen wells. In some of the cases we have documented, shareholders with landholdings of different sizes disagree on the distribution of benefits from well-deepening, and one or more shareholders refuse to contribute to the cost. Conflicts arising out of such circumstances are again referred to the Panchayats. The Panchayats often "solve" such disputes by dividing wells physically into as many shares as needed, thereby placing the responsibility on the individual shareholders to dig and deepen their delineated parts. Such physically fragmented wells are common in all the villages surveyed. Although this approach is common, it often encourages competitive deepening among shareholders of individual wells, effectively leading to the construction of wells within wells. In such cases, shareholders lacking the resources to deepen their own portion lose access to groundwater and the well is effectively controlled by those who afford to deepen them to extract more water. There are also instances of wells being abandoned due to multiplicity of shareholders and recurrence of numerous disputes.

The history of each jointly owned well surveyed in the Palar basin was recorded. Initially, most wells were individually owned. Division into shares occurred subsequently, due primarily to the operation of inheritance laws. When land is divided among legal heirs, wells are also divided. Therefore, most shareholders in jointly owned wells are brothers or close cousins. Over time, however, shares are often sold to others for many socio-economic reasons. In a few cases, single well-owners have approached neighbors to share the cost of well-deepening and effectively sold their shares in the wells.

While sharing of water from a jointly owned well is often problematic, positive features of this sharing also exist. The fact that at least one-third of the wells in our survey areas were jointly owned indicates the sustainability of this system. Indeed, in all the villages, there were institutionalized (informal) rules governing the sharing of water from jointly owned wells. Ostensibly, the joint well system promotes the equitable use of groundwater and particularly benefits those who cannot afford a well of their own. Many jointly owned wells however, fail for two interrelated reasons: declining groundwater levels and lack of funds for well-deepening. Thus, many joint well-owners become heavily indebted and are eventually forced to sell their shares along with their portions of landholding. Though the sharing of a well would seem to promote equitable use of groundwater, eventually it only reinforces inequity.

Ownership of Wells Across Size Categories of Farmers

Because the development of a well for irrigation requires substantial investment, it is often portrayed as affordable only for resource-rich farmers. Our data, however, does not support this perception. The data collected from the survey of 27 villages in the Vaigai basin indicate that nearly three-fourth of wells are owned by farmers owning 5 acres or less (Janakarajan 1997a). A similar survey of eight villages of the Palar basin indicates that the 65 percent of farmers whose landholding size is less than or equal to 4 acres own 54 percent of all wells. This group, however, owns only 29 percent of the total land held by the farmers interviewed. The average area irrigated per well is 1.46 acres in this size class. In contrast, the 3 percent of farmers owning more than 15 acres also own 8 percent of the wells surveyed and 19 percent of the total land. The average area irrigated by a well in this size class is 26 acres. More detailed data are given in Table 7.5 and Figures 7.4 and 7.5. These data indicate that, while the wealthy do tend to own more wells, the distribution is far less skewed than is land ownership. Average well-ownership per unit of land, in fact, declines exponentially as landownership size increases. The data do not, however, indicate the type and productivity of the wells owned by different size classes of farmers. Since the average area irrigated per well is far larger in the big landholding classes, the wells may be more productive and actual access to groundwater may be more skewed than suggested by comparisons between well- and land-ownership alone.

Although the data indicate that well-ownership is far less skewed than landownership, a number of factors suggest that the poor may

Table 7.5
Ownership of wells across size classes of landholding in Palar basin

Landholding Size (acres)	*No. of Well-owners*	*Total No. of Wells Owned*	*Total Extent of Land Owned/ Irrigated (acres)*	*Average Extent of Land Irrigated per Well (acres)*
Less than 1	26	29	16.7	0.64
1.01–2	64	86	101.7	1.59
2.01–4	67	100	193.9	2.89
4.01–6	28	43	140.8	5.03
6.01–10	35	75	257.7	7.36
10.01–15	14	35	173.8	12.42
15.01–25	5	13	97	19.40
Over 25	3	17	111.1	37.04
Total	242	398	1092.7	4.52

Source: Main survey (1998–2000).

Figure 7.4
Well- and land-ownership in Palar basin

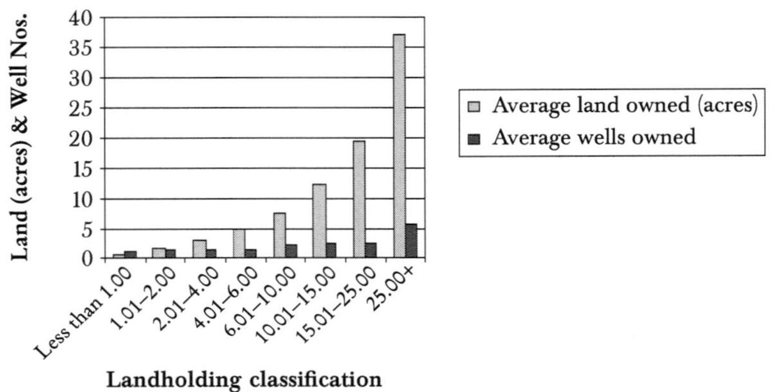

Source: Main survey (1998–2000).

Figure 7.5
Average number of wells owned per unit of land in Palar basin

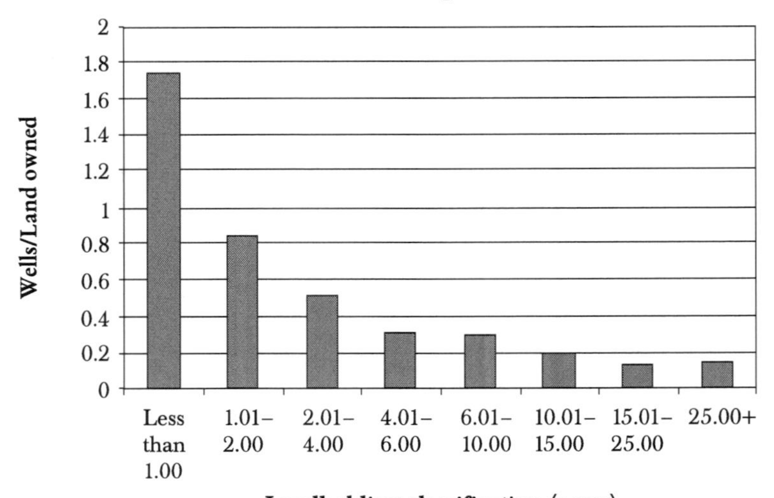

Source: Main survey (1998–2000).

not be deriving as much benefit as it appears. First, as the data in Table 7.5 on area irrigated per well indicate, wells owned by farmers with big landholdings are likely to be much more productive and capable of irrigating larger areas than are wells owned by farmers with smaller landholdings. Second, as groundwater levels decline, large farmers are able to devote more resources to deepening their wells than are small farmers. In addition, access to larger land areas implies access to a wider variety of potential sites for digging wells. Because hard-rock geology is highly variable, access to a variety of locations for digging new wells is often critical to agricultural success. Third, many of the more wealthy farmers established wells earlier than did the small farmers and were able to benefit from them even before competitive deepening of wells became a major issue. As a result, although the poor appear to own large numbers of wells, many are trapped in a regime in which water table is retreating progressively. Their position is, therefore, quite vulnerable. In order to be able to remain in the race for well-deepening, they have to keep investing in well-deepening without any assurance of accessing substantial volumes of groundwater. While some are successful, the majority fail and are pushed into a debt trap. We will get back to this issue in a later section.

Linkage between Surface and Groundwater

Extensive development of groundwater resources is affecting surface water systems in the Palar basin. The Palar basin is known for its rich River Bed Aquifer (RBA) which contributes substantially to the "spring" channels and, although extraction is prohibited, to thousands of wells located along the riverbed. Pumping of groundwater in the prohibited areas is drying up surface water bodies and resulting in reduced downstream flows. Over 100 million liters per day (mld) of water is pumped from the Palar riverbed for drinking and industrial purposes. Although the volume of water pumped for industrial and domestic purposes is smaller than what is pumped for irrigation, it has adverse effects in two ways: one, what is pumped for domestic and industrial purposes is of a potable quality, which is not available in all the villages, and two, such extraction of groundwater takes place in some select regions or villages, causing tremendous stress on the local economy. Furthermore, this has a direct impact on the traditional "spring" channels which were originally constructed to tap sub-surface flows in the river. These "spring" channels traditionally provided water for irrigation for at least one full cropping season. Historically, at least one such "spring" channel provided water for each village located along the river course. Thousands of such "spring" channels are reported to have existed in Tamil Nadu as per the village records. Most of these have now dried up and have been encroached upon. Out of the 51 villages surveyed in the Palar basin, "spring" channels were found to be practically defunct in 35, functioning poorly in six, and fairly effective in only three. In the remaining villages, "spring" channels were found to have been taken over by the tanneries for discharging industrial effluents. Since these channels carrying industrial effluents passed through interior parts of the villages, even groundwater was heavily polluted.

In addition to having an adverse impact on RBAs, unregulated pumping of groundwater in tank commands has also had serious consequences. Since a significant number of wells are located in tank commands, the tank is losing its place as an important source of irrigation (Vaidyanathan and Janakarajan 1989). The rapid spread of well irrigation, accompanied by large-scale rural electrification and introduction of high-yielding variety (HYV) of crops have contributed in a great measure to the rise of conflicting interests in the use of ground and surface water. Since high-yielding varieties required more assured, controlled and timely application of water and since the available tank water is inadequate to raise three short-duration HYV crops, wells have major advantages over surface water sources. Furthermore, some studies indicate a positive correlation between the rapid growth of well irrigation and the decline of traditional irrigation systems, such as tanks (Palanisamy et al. 1996; Janakarajan 1993; Vaidyanathan and Janakarajan 1989). Lindberg (1999) in his paper shows how individual rationality conflicts with collective rationality and eventually results in the erosion of common property resources (CPRs). Individuals have strong incentives to disassociate themselves from collective tank maintenance and pump groundwater indiscriminately. This results in progressive lowering of the water table. The government's policy of supplying free electricity to agriculturists has aggravated this problem. All this leads to general environmental degradation, high groundwater extraction and decline in aquifer recharge due to the drying up of the surface water bodies, such as tanks.

In our survey, traditional irrigation systems were found to be defunct in six out of the 17 tank commands studied in the Palar Anicut System. These were also the tank commands in which well density (i.e., number of wells per acre) was quite high. In one of the tanks, the tank sluices were kept closed permanently to facilitate the recharge of wells located in the tank commands. In the rest of the operational tanks, the traditional irrigation system was reasonably unimpaired, but these were also the tank commands in which the well density was very low (Janakarajan 1993; Vaidyanathan and Janakarajan 1989). Similar finds were made in a large-scale study undertaken by Tamil Nadu Agricultural University (TNAU) (Palanisamy et al. 1996). The close association between a high well density and disintegration of tank irrigation systems has also been found in other village-level studies carried out in Tamil Nadu (Janakarajan 1997b, 1986; Harriss 1982; Nanjamma 1977). The findings are, however, not uniform. A study of tanks in the Periyar–Vaigai system shows that the spread of well irrigation in the tank commands did not lead to a total collapse of the tank irrigation system although its degree of effectiveness varied according to the well density (Vaidyanathan and Sivasubramanian 2004).

Our recent survey in 51 villages of the Palar basin indicates that there exists a close association between well density in the command area of tanks and springs, and the decline of these traditional sources of irrigation. In the villages surveyed, well density ranged from a low of 0.30 to a high of 0.79 per hectare; well density in wet lands, traditionally irrigated from surface water sources, were found to be typically higher (0.33–0.79 wells per hectare) than that in dry lands (0.30–0.62 wells per hectare). This density was much higher than expected even in villages where tank irrigation institutions are reported to remain alive. According to interviews with farmers, the dependability of tank water was low and the risk and uncertainty associated with relying on it high. As a result, many farmers invested in wells to get access to more assured irrigation water. The tanks, if they functioned at all, were used as percolation ponds in most of these villages. Indeed, access to private source of irrigation water (wells) provided generous disincentive to farmers for non-cooperation in the collective maintenance of tanks and "spring" channels.

At one level, it can be argued that pumping recharged groundwater is a more efficient way of using water than practicing surface irrigation. In fact, in several villages, better-off farmers (multiple well-owners) find it convenient and useful to close down the sluices of tanks so that the impounded tank water constantly recharges their wells. But, in many cases, since there is absolutely no maintenance of inlet channels, tanks and springs are heavily silted and store very little water. This has serious implications for non-well-owners who were solely dependent upon tank water.

Pollution, Cropping Pattern and Yield

The Palar and Noyyal river basins are under severe stress not only due to over-use of groundwater but also due to pollution. It is, therefore, necessary to analyze irrigated areas, crop patterns and crop yields in these basins.

In the main survey of the Palar and Noyyal basins, the net irrigated area per well in villages where groundwater has been affected by pollution was found to be 2.72 acres; the average net irrigated area in the areas where groundwater had not been polluted was found to be 4.16 acres. Differences in cropping patterns are even more striking. The total area for all crops grown on land irrigated by sample wells in the villages surveyed in the Palar basin was 903 acres, of which 505 acres (56 percent) was devoted to paddy. Over 90 percent (456 out of 595 acres) of this paddy was grown in villages where groundwater had not been polluted. This is equivalent to 2.9 acres of irrigated paddy per sample well in villages with unpolluted groundwater and only 0.50 acres of irrigated paddy per sample well in villages with polluted groundwater. The cropping pattern in villages with polluted groundwater was that larger areas were devoted to the cultivation of sugarcane and coconut, crops that tolerate polluted groundwater reasonably well (see Figure 7.6). Differences in the cropping patterns were not as great in the Noyyal basin because paddy is not a major crop there.

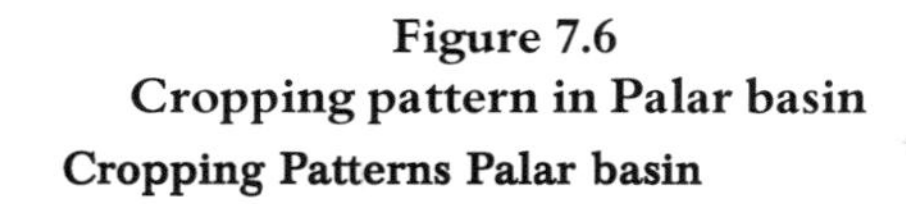

Figure 7.6
Cropping pattern in Palar basin

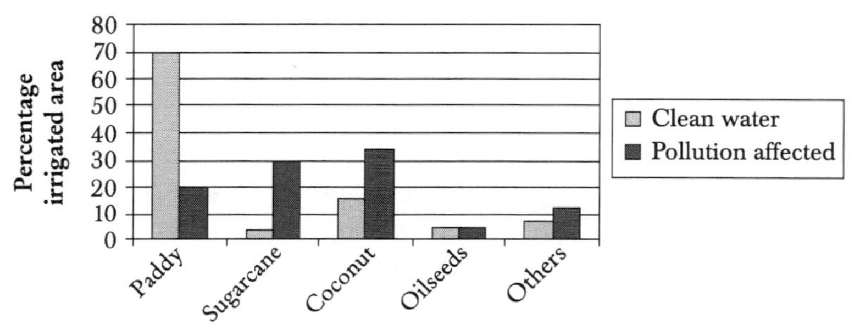

Source: Main Survey (1998–2000).

The contribution of groundwater over-use and pollution to water scarcity is major in both Palar and Noyyal basins. In about 33 percent (in 80 out of 253 sample wells) and 28 percent (80 out of 253) of sample wells in Palar and the Noyyal river basins respectively, irrigated area was nil, implying that wells were no longer utilized. The difference between the irrigated areas in villages with polluted groundwater and those in villages with unpolluted groundwater is substantial: on the one hand, in Palar river basin, 26 percent of wells (41 out of 159 sample wells) in villages without polluted groundwater and 41 percent of wells (39 out of 94) in villages with polluted groundwater had no irrigated area; on the other hand, Noyyal river basin, 25 percent of wells (28 out of 112 sample wells) in villages with unpolluted groundwater and 34 percent of wells (23 out of 68 sample wells) in villages with polluted groundwater reported zero irrigated area.

Differences between the net area irrigated by wells in villages with polluted groundwater and that in villages with unpolluted groundwater have a large impact on crop yields. About one-third of the sample well-owning farmers in both the river basins reported zero crop yield. Again, the difference between villages with polluted groundwater and those with unpolluted groundwater is substantial. While in the villages with polluted groundwater, 43 percent of the sample well-owning farmers reported zero crop yield, only 28 percent did so in villages with unpolluted groundwater. In both types of villages, however, the incidence of sample well-owning farmers reporting zero crop yield is quite significant. In the case of villages where groundwater had not been polluted, zero crop yield was caused by groundwater over-extraction and the drying up of wells. In the villages where groundwater had been polluted, zero crop yield was primarily due to severe water contamination. The economic impact of pollution are also evident in the value of crop production in different villages. For instance, 79 out of 159 sample wells (50 percent) in the villages of Palar basin with unpolluted groundwater and 60 out of 112 sample wells (54 percent) in the villages with unpolluted groundwater in Noyyal basin reported more than INR 5,000 worth of crop yield per acre. In contrast, only 16 out of 94 sample wells (17 percent) in the pollution-affected villages of Palar basin, and 11 out of 68 sample wells (16 percent) in the pollution-affected villages of Noyyal reported more than INR 5,000 worth of crop yields per acre. These impacts are particularly important for small farmers who cannot deepen wells or dig new wells in less polluted locations. As the accompanying box illustrates (see Box 1), however, the impact of pollution even on large farmers is often very substantial.

Decline in the Water Table, Competitive Deepening and its Socio-economic Implications

In many parts of India, rapid expansion of groundwater irrigation has resulted in a serious decline in groundwater levels; in some cases, pumping rates exceed rate of recharge resulting in groundwater mining (see, for instance, Janakarajan 1997a; Vaidyanathan 1996; Rao 1993; Bhatia 1992; Moench 1992a). This is widely viewed as a major cause of competitive deepening and emergence of conflicting interests among well-owners. Little data are, however, commonly available on the extent to which decline in groundwater level has actually occurred in specific locations. The most recent formal statement on the status of groundwater resources in India by the Central Ground Water Board (CGWB) was published in 1995 and is based primarily on data from the period 1989–90 and contains no information on actual changes in groundwater levels. Furthermore, in most states, data on groundwater monitoring are insufficient to accurately capture changes in groundwater levels

Box 1
Noyyal Basin; Village: Orathapalayam;
Sample Well Code No.: OPM

This well-owner has five wells and 18 acres of land. All the wells are interconnected with pipelines. His original objective was to pump water from all the wells, channel them together for irrigation. This arrangement was done because the yield of water from his wells was low. The wells ranged from 50 to 70 feet in depth and the total amount spent in constructing them and installing five pumps, pump sheds, pipelines and other equipments came to INR 1.3 million (approximately US$ 27,000 at the exchange rate of US$ 1: INR 48). The farmer profitably engaged in agriculture until the late 1980s. In 1990, a dam was constructed in this village across Noyyal river to irrigate 11,000 acres of land. The well-owner's destiny has changed since then. The dam collects all the effluent water discharged by 750 dyeing and bleaching units located in and around Tiruppur town. Because of very high presence of Total Dissolved Solids (TDS), other chemicals and salts in the stored water, it was never been used for irrigation. Unfortunately, however, all the wells belonging to the farmer were adjacent to the dam and became polluted. This farmer is at present growing coconuts which tolerate salinity to some extent. His annual income has declined from about INR 0.3 million (US$ 6,250) to less than INR 50,000 (US$ 1,042). He has accumulated debts amounting to INR 0.4 million (US$ 8,333). The condition of many small well-owning farmers is much worse; they have given up farming in this village and sought employment in the Tiruppur knit-wear, dyeing and bleaching industries.

at a local level even if the data were made generally available (Janakarajan 2001; World Bank 1998; Moench 1994).

Given the lack of detailed monitoring data, our approach to estimating changes in groundwater levels in the study areas was to collect through survey information on the original and current depths of sample wells. These data indicate that decline in groundwater levels have been significant both within and outside the canal and tank command areas.

Declining groundwater levels are clearly indicated by the changes in original and current depths of wells in both Palar and Noyyal basins. These data are presented in the two figures (7.7 and 7.8) and are based on a survey of 237 wells in eight villages of Palar basin and 171 wells in four villages of Noyyal basin, conducted between 1998 and 2000. The data are about wells located in both dry and wet lands. The data show that wells in both basins have been deepened over time. The increase in depth is particularly

Figure 7.7
Change in the original and the current well depths in Palar basin

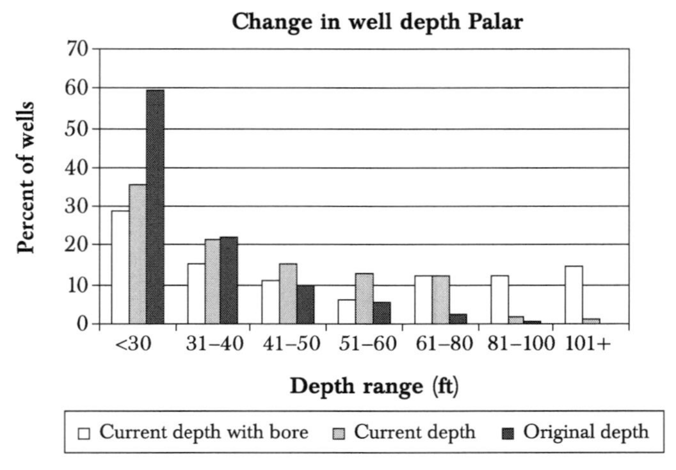

Source: Main survey (1998–2000).

Figure 7.8
Change in the original and the current well depths in Noyyal basin

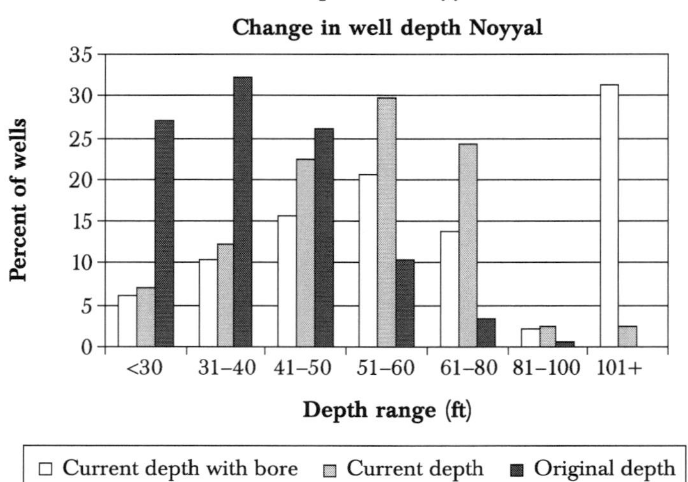

Source: Main survey (1998–2000).

pronounced if the bores drilled within dug wells are included. In the Palar basin, almost 60 percent of wells were initially less than 30 ft deep, but now less than 30 percent are so, if the depth of bores is factored in. Originally, no well was more than 100 ft deep, now over 14 percent are. The change is even more dramatic in the Noyyal basin where originally almost 60 percent of wells were less than 40 ft and now only 17 percent are; and moreover, more than 30 percent exceed 100 ft in depth.

Box 2
Palar Basin; Village: Kathiavadi;
Sample Well Code No.: KYD 40

This well was dug in 1938 to a depth of 15 ft. Between 1950 and 1985, it was deepened six times to a total depth of 39 ft. An electric pump replaced the manual lift during the mid-1960s. There are three adjacent wells located within a radius of 150 ft, whose depths were initially around 30 ft. A series of droughts in the late 1980s made all the well-owners deepen their wells. By 1992, the depth of the sample well was 50 ft and it had both vertical and horizontal bores installed within it. This caused two of the adjacent wells to dry up and reduced yields in the third. Their owners now lack the financial resources to deepen their wells any further. In contrast, the owner of the sample well is irrigating about 5 acres of his own land and is selling water to the others so that they can irrigate another 2 acres in each season. This is the clear case of competitive deepening where one well-owner has been able to maintain or increase his prosperity while others are forced to purchasing water.

In addition to the overall decline in groundwater levels, earlier studies have indicated that the "original depth" to which wells needed to be dug has increased over time: a "new comer" has dug deeper than his predecessor had to, say 10 years ago (see Janakarajan 1997a). This is confirmed by data from the current survey. In the Palar basin, the average original depth of the sample wells dug before 1960 was 30.2 ft. It rose to 35.8 ft for wells dug between 1961 and 1970, to 41 ft for wells dug between 1971 and 1985, and and has averaged 69 ft for all wells dug subsequently. Similarly, in the Noyyal basin, the average original depth of sample wells dug before 1960 was 42.6 ft, while the depth of wells dug after 1985 averages 66 ft. If one includes bore wells (which are more common in the Noyyal than in the Palar basin), the depth has increased from 100 ft between 1960 and 1970, the period in which the first bores were installed, to 260 ft in the post-1985 period. For the Noyyal basin, this suggests that the annual rate of decline in groundwater level is approximately 10 ft.

Changes in the well depth have been accompanied by changes in the water lifting technologies. In the Palar basin, of the 253 sample wells surveyed, 191 reported the past use of *kavalai* (bullock bailing lift) as the original technology, only one of which was still reported to exist at the time of the survey (that too not in use). Similarly, in the Noyyal basin, out of 181 sample wells, 121 wells reported *kavalai* as the original technology, while only one was operational at the time of the survey. The disuse of *kavalai* probably owed to two factors: decline in groundwater levels which reduced the functionality of manual lift devices within dug wells, and the spread of mechanized pumping technologies, combined with and the necessity of using mechanical pumps in bore wells. It is also interesting to note that the number of wells with no water lifting device (WLD) has gone up considerably over time, from three (as per the original WLD) to 71 (as per the current WLD) in the Palar basin and none to 19 in the Noyyal basin. These are the wells that have been deepened but subsequently abandoned due to either lack of water supply or bad water quality.

Box 3
Noyyal Basin; Village: SA Palayam;
Sample Well Code No.: SAP 2

This well-owner initially had an open well and used it until 1980. This well was 70 feet deep with six vertical and six side bores. The well stopped yielding water during a drought in the 1980s despite an investment of over INR 0.3 million. It was permanently abandoned in 1990 when neighbors installed 250 feet deep bore-wells. At that point, he also decided to install deep bores. Over the last 10 years, he has installed 10 bores in different parts of his land to depths ranging between 300 and 700 feet. Out of these, only two, one of which is the deepest, supply water and more than 25 borewells around his well have dried up. He has spent INR 0.5 million on all these bores and can now cultivate coconut and tobacco on 8 acres out of his total landholding of 20 acres. As a rich farmer who also owns a tobacco processing company, he has no debts. His income is, however, derived primarily from the tobacco company (which employs 100 women), not farming. During our interview, he proudly informed us that his neighboring farmers decided to sell their lands because their bores had dried up.

As illustrated in the accompanying Boxes 2 and 3, declining groundwater levels have led to extensive competition between well-owners. The vast majority of farmers have deepened their wells several times. In addition, because many farmers have installed horizontal as well as vertical bores, the impact on water availability in adjacent wells is often severely affected. While disputes over water and the deepening of wells are common, no dispute was reported in our survey due to side-bore installations even when they penetrated adjacent lands. Despite the extent of competition and conflict over deepening of wells, farmers do not seek justice through the court of law because property rights in groundwater are known to be ambiguous and indeterminate. This situation has serious negative implications for future users and adds tremendously to the costs faced by the current users (Janakarajan 1997b). Finally, it is important to mention before concluding this section that competitive deepening is virtually absent in pollution-affected villages, since farmers do not have incentive even to use groundwater for irrigation.

Impact of Decline in Groundwater Levels on Well Technology

Falling groundwater levels and competition have serious implications for the forms of well technology that can be used. First, there has been a change in the design and type of wells dug. Conventional, large-diameter, round or square wells cannot be used when groundwater levels fall; hence, new technologies for both well-digging and pumping have spread in recent decades. Now, a large majority of wells in the Palar river basin are fitted with both vertical and horizontal bores, and in the Noyyal most farmers install deep bores from the surface. Hydraulic drilling companies have mushroomed in the Noyyal region and are making huge profits. This kind of well-digging technology has substantially contributed to competitiveness and over-pumping of groundwater.

Second, well-deepening and use of high-power motors and compressors have had a huge impact on energy demand. Until three decades ago, bullock bailing was the main method of groundwater extraction, a practice that is almost extinct now. It was followed by pumping of groundwater with low-capacity (3.5 HP) pumpsets, a practice that continued till mid-1980s. Now, use of 10 HP motors is common, particularly in the Noyyal region, and in many cases farmers use more than one motor to pump water from the same well. All this has been facilitated by the state's policy of providing free power supply.

Third, declining groundwater levels have encouraged increase in efficiency of water use. Until the late 1980s, open channels were used for conveying water from wells to fields. Now, farmers often use underground pipelines and hose pipes.

Fourth, high well-digging and equipment costs disproportionately affect small farmers who own about 60 percent of wells in the state. While big farmers have the resources to bear unsuccessful investments in well-digging and well-deepening or persistent droughts (as had occurred in the 1980s), for small farmers, such losses are often unsustainable.

It is worthwhile looking at the impact of competition on changing technologies in more detail. The case of the Noyyal illustrates the ongoing changes well. Unlike the situation in Palar basin, in the Noyyal basin groundwater is extracted from deep bores. In some locations in Noyyal, depth of borewells approaches 1,200 ft. Due to the hard-rock nature of the area's geology, yields in such wells are very low, making continuous pumping difficult or impossible. Wells need time to recuperate – for gradual seepage from fractures in the bedrock to re-establish a water column – before water can be pumped from them again. To assist this process, farmers use compressor technology which allows them to run pumps even when there is very little water. Approximately 95 percent of the borewells in this basin are fitted with compressors. With compressors, the amount of water that can normally be pumped in one hour takes six to seven hours. Since the yield is low, the flow is insufficient for irrigation or sale. As a result, water is pumped and stored in cisterns that are either adjacent dry dug wells or concrete tanks of up to 100, 000 liter capacity. From these cisterns, water is pumped again for irrigation or sale. The electricity consumption in these borewells is double or triple due to (*a*) the use of compressors to run pump motors, (*b*) the running of motors for long periods of time to pump small amounts of water, and (*c*) the need to pump the same amount of water twice (once from the bore and again from the open well tank where the water is stored).

Because of low yields and compressor technologies, the way water is pumped and stored has major implications for both energy use and the overall cost of accessing groundwater. On the basis of sample survey data collected in four villages of this basin, we have developed a typology that illustrates the diverse techniques and equipments required (see Table 7.6). As Table 7.6 demonstrates, farmers often need to invest in high-capacity pumps and in the substantial storage structures. Low yields also often require farmers to drill multiple bores within dug wells. Finally, in many cases (37 percent of the sample wells) the same amount of water is pumped twice, once directly from the well and again for irrigation or sale.

Costs of and Investments in Well-digging

The variety of pumping and well technologies now in use has major implications for the cost of obtaining access to groundwater. The cost of digging a well is much lower in the Palar basin than in the Noyyal because water tables are higher and the more expensive compressor and storage technologies are not required. In the Palar basin, the average cost of pumping equipment is INR 14,600 per well (including motors, pumps and other related accessories). In the Noyyal basin, equipment costs average INR 31,000. In addition, each successfully operating borewell requires at least five to six trial-bores. Furthermore, around each operating bore or open well, there are several closed bore points which have stopped yielding water. There is, however, no assurance that successful wells will remain productive. Indeed, according to the latest available statistics for Tamil Nadu (GoTN 1998), the wells that are not in use constitute about 10 percent of the total number of wells in the state. Further, many wells have been abandoned even after investing over INR 100,000. (Janakarajan, 1997a). Eventually, all the investments that have gone into wells cumulatively put a heavy burden on the community as a whole as well as on an

Table 7.6
Typology of wells with different pumping and storage equipments

Well, Pump and Cistern Characteristics	*No. of Wells in Village**			
	Kar	*Ora*	*Sou*	*Uga*
Deep borewell from which water is pumped with one motor and one compressor in order to store water in an independent well – only to pump again for irrigation (twice pumped)	8	2	16	15
Deep borewell from which water is pumped with one motor and one compressor in order to store water in an independent well – only to pump again for water sale for industries (twice pumped)	0	0	0	3
Deep multiple bores (up to 3) simultaneously operated with one high-power motor (10 HP), along with one compressor, in order to store water in a deep open well – only to pump again for irrigation (twice pumped)	5	0	3	8
Deep multiple bores (more than 3) simultaneously operated with two high-power motors (of up to 10 HP each), along with two compressors, in order to store water in a open well – only to pump again for irrigation (twice pumped)	1	0	0	0
Shallow well which is operated (with up to 5 HP motor) for direct irrigation – own use (once pumped)	6	5	6	7
Shallow well which is operated (with up to 5 HP motor) for sale of water to industries and for irrigation (once pumped)	0	0	0	2

Deep well which is operated (with up to 7.5 HP motor) for direct irrigation – own use (once pumped)	5	7	18	10
Deep multiple (vertical bores installed within a dug well) operated with a high-power motors – used for own agriculture – (once pumped)	5	0	2	4
Multiple deep bores (up to 4) from which water is pumped simultaneously with a single high-power motor (of up to 10 HP) with one compressor in order to store water in a concrete tank (with a capacity of 100,000 liters) both to sell water for urban industrial use and pump for own agricultural use (twice pumped)	2	0	0	1
Multiple deep bores (up to 4) from which water is pumped simultaneously with a single high-power motor (of up to 10 HP) with one compressor in order to store water in a concrete tank (with a capacity of 100,000 liters) but used for dyeing/bleaching [D/B] industries), (twice pumped)	3	0	0	0
Multiple deep bores (up to 4) from which water is pumped simultaneously with a single high-power motor (of up to 10 HP) with one compressor in order to store water in a concrete tank (with a capacity of 100,000 liters), but water is used for own agricultural use by letting water through gravity flow (once pumped)	1	0	0	1
Shallow wells, not in use	12	6	9	8
Total of all types	48	20	54	59

Source: Main survey (1998–2000).

Note: Village *names:* Kar = Karaipudur; Ora = Orathapalayam; Sou = South Avinashipalayam; Uga = UGAYANUR.

individual farmer. The cost is not, however, just at the community level. Since electricity for agricultural pumpsets is free in Tamil Nadu, this cost is paid by the tax-payers as a whole. Farmers face no marginal cost and do not hesitate to pump water even if the delivery of water is quite low.

As a part of the survey in the Palar and Noyyal river basins, we collected basic information on the investments farmers have made to first get and subsequently maintain access to groundwater. These data are discussed later in the chapter. Before presenting the results, it is important to note the key limitations of the data. In most instances, the figures well-owners gave are below the current prices. As a result, the current value is likely to be higher than the data suggest. In addition, significant difficulties were faced in gathering this information due to memory lapses, sale, inheritance and transfer of wells to others. In consequence, data for some sample wells are not included in our analysis.

Our data indicate that the cost incurred on wells by individual farmers is high and and often disproportionate to the level of farm income generated. In addition, it varies from wet lands (those located in the command area of surface water systems) to dry lands. The amount spent per well in wet and dry lands of the Palar basin (aggregate for eight sample villages) is INR 72,000 and INR 86,000 respectively at current prices. Wells in wet lands tend, however, to supply water to much larger command areas and require lower investments in supplementary equipments. As a result, although the well costs differ by less than 20 percent, the net costs are 36 percent higher in dry lands (INR 70,000 per hectare) than in wet lands (INR 95,500 per hectare). Costs incurred for wells in dry land are much higher because the water table has declined much more steeply than in the wet lands. In the Noyyal river basin, average current cost per well (INR 221,000) and the cost incurred per hectare of net irrigated area (INR 188,000) are higher than in the Palar basin (see Tables 7.7 and 7.8).

Two points are worth noting from Tables 7.7 and 7.8. First, the costs incurred per well and per hectare are high. According to the Ninth Five Year Plan Document (1997–2002), the cost incurred to create one hectare of major and medium irrigation potential by the government works is INR 40,166 at current prices. (GoI n.d.). In our survey, individual farmers were found to spend INR

Table 7.7
Cost of well irrigation in wet- and dry-land wells of Palar basin

Village	*No. of Sample Wells*	*Original Cost Per Well (INR)*	*Average Current Cost Per Well (INR)*	*Original Average Cost Per Hectare of NIA (INR)*	*Current Average Cost Per Hectare of NIA (INR)*
Kathiavadi	13	2,615	91,000	1,935	67,000
Poondi	15	8,733	79,000	6,488	58,000
Gudimallur	7	857	86,000	534	54,000
Periavarigam	5	8,800	58,000	9,205	61,000
Solur	5	1,800	51,000	5,556	159,000
Damal	38	13,289	75,000	4,297	24,000
R. N. Pettai	8	4,875	65,000	7,800	104,000
N. M. Pattu	8	6,250	87,000	2,317	32,000
Average		8,242	72,286	4,767	69,875
Cost of well irrigation in the dry-land wells in Palar basin					
Kathiavadi	27	11,074	116,000	8,413	88,000
Poondi	7	16,857	84,000	19,250	96,000
Gudimallur	12	5,583	79,000	9,293	131,000
Periavarigam	25	5,400	93,000	6,139	105,000
Solur	16	7,063	93,000	7,766	103,000
Damal	11	16,000	81,000	6,780	35,000
R. N. Pettai	34	10,471	76,000	19,734	143,000
N. M. Pattu	18	10,444	68,000	8,835	58,000
Average		10,362	86,250	10,776	94,875

Source: Main survey (1998–2000).

Note: Groundwater in Solur, Periavarigam, Gudimallur and Poondi is heavily polluted by tannery effluents; groundwater in Kathiavadi is partially polluted; and groundwater in Damal, N. M. Pattu and R. N. Pettai is not polluted.

Table 7.8
Cost of well irrigation in Noyyal basin

Village	*No. of Sample Wells*	*Original Cost per Well (INR)*	*Average Current Cost per Well (INR)*	*Original Average Cost per Hectare of NIA (INR)*	*Current Average Cost per Hectare of NIA (INR)*
S. A. Palayam	54	9,907	230,333	7,778	180,837
Ugayanur	59	22,797	199,559	20,345	178,097
O. Palayam	20	9,000	202,450	6,020	135,418
K. Pudur	48	21,000	252,521	21,279	255,879
Average	–	15,676	221,216	13,856	187,558

Source: Main survey (1998–2000).
Note: Groundwater in O. Palayam and K. Pudur villages is heavily polluted by effluents from the dyeing and bleaching industries; groundwater in other two villages, viz., S. A. Palayam and Ugayanur is not polluted.

70,000 and INR 95,000 to get one hectare of net area irrigated by wells in the wet and dry lands respectively in the Palar basin. In the Noyyal basin, the costs per hectare (average INR 190,000) are far higher – approximately 4.7 times what the government has spent to create one hectare of irrigation potential under major and medium irrigation projects. Newcomers would need to spend this much to develop one hectare of net area irrigated by wells. In addition, they have to bear the risk of failures due to decline in groundwater levels, droughts or problems in locating a productive zone. Further, there is substantial variation between villages. Local groundwater conditions and the presence or absence of pollution have a huge impact on the costs of digging or deepening wells and on the irrigated area. Well irrigation has thus become a gamble. Not all those who invest in wells are successful. Many lose in the race of competitive deepening, or wells go into disuse due to pollution. Many well-owners either sell their land or become indebted in trying to dig new wells. A new dimension of inequality emerges as a result of this. Those who have, so far, been able to stay ahead in the race for competitive deepening have emerged as potential water sellers, while others have been reduced to the status of water purchasers (Janakarajan 1997b; Vaidyanathan 1996).

Water Markets: Conflicts and Contradictions

As the cost of digging or deepening wells has increased, the sale of groundwater in rural areas has become a common phenomenon. Like joint well-ownership, the emergence of water markets in rural areas is a spontaneous institutional response to scarcity, which facilitates sharing of scarce groundwater resources. The magnitude of water markets and the terms and conditions under which they operate vary greatly depending upon the availability of groundwater, water quality, soil conditions and a variety of other factors. While a full review of water markets is beyond the scope of this chapter, a number of points are important to note.

First, the price paid for water is often dictated by the nature of the water supplier. If the state is the water seller, the price individuals are willing to pay is insignificant in India compared to what is paid to a private seller. As the Committee on Pricing of Irrigation Water reports, "At present, the actual gross receipts per hectare of area irrigated by major and medium projects is barely 2 per cent of the estimated gross output per hectare of irrigated area"(Planning Commission, Government of India, 1992). On the other hand, farmers pay up to one-third of their gross produce or up to INR 40 per hour for water supplied by a private well-owner (Janakarajan 1997a, 1992).

Second, private water sellers pay very little or nothing for power to the state particularly in Tamil Nadu and Haryana. As Table 7.9 indicates, electricity tariffs in Tamil Nadu were first subsidized for small farmers, then charged on a horsepower basis from the early 1980s. Since 1989–90, power for agricultural users has been supplied free of cost. Power consumption in agriculture has, as a result, been increasing. This is not, however, due primarily to the growing number of pumps. Although the number of pumps has been increasing, the time series data clearly indicate that power consumption per pumpset has been increasing at a faster rate than the total increase in agricultural energy consumption. This change is particularly evident from the year free electricity was introduced (see Figure 7.9). This may be due to long-term decline in the water

Table 7.9

Electricity tariff and consumption of electricity per pumpset in Tamil Nadu, from 1970–71 to 1996–97

Year	*Total Energy Consumed for Agricultural Pumpsets (million units [mu])*	*No. of Electric Pumpsets*	*Energy Consumed/ Pumpset (units)*	*Tariff Charged for Agricultural Pumpsets*
1970–71	1,241	529,932	2,342	8 paise/unit
1971–72	1,269	594,169	2,136	9 paise/unit
1972–73	1,430	649,241	2,203	11 paise/unit
1973–74	1,576	681,205	2,314	11 paise/unit
1974–75	1,847	706,914	2,613	11 paise/unit
1975–76	1,675	742,745	2,255	16 paise/unit
1976–77	1,697	773,702	2,193	16 paise/unit
1977–78	1,786	809,606	2,206	Big farmers: 16 paise/unit Small farmers: 14 paise/unit
1978–79	2,104	840,557	2,503	Big farmers: 14 paise/unit Small farmers: 12 paise/unit
1979–80	2,186	887,227	2,464	Big farmers: 14 paise/unit Small farmers: 12 paise/unit
1980–81	2,299	919,162	2,501	Big farmers: 14 paise/unit Small farmers: 12 paise/unit
1981–82	2,354	945,520	2,490	Big farmers: 15 paise/unit Small farmers: 12 paise/unit
1982–83	2,230	965,017	2,311	Big farmers: 15 paise/unit Small farmers: 12 paise/unit

1983–84	2,200	982,606	2,239	Big farmers: 15 paise/unit Small farmers: 12 paise/unit
1984–85	2,415	982,606	2,458	Big farmers: INR 75/HP/year Small farmers: INR 50/HP/year
1985–86	2,840	1033,533	2,748	Big farmers: INR 75/HP/year Small farmers: INR 50/HP/year
1986–87	3,114	1074,184	2,899	Big farmers:INR 75/HP/year Small farmers: INR 50/HP/year
1987–88	3,136	1,116,177	2,810	Big farmers: INR 75/HP/year Small farmers : INR 50/HP/year
1988–89	3,524	1,184,450	2,975	Big farmers:INR 75/HP/year Small farmers: INR 50/HP/year
1989–90	3,740	1,235,941	3,026	Big Farmers:INR 75/HP/year Small farmers: INR 50/HP/year
1990–91	3,974	1,318,671	3,014	INR 50/HP/year for £ 10 HP and INR 75/HP/year for >10 HP
1991–92	4,451	1,359,748	3,273	Since 1991 free supply for all
1992–93	5,160	1,403,673	3,676	
1993–94	5,618	14,45,951	3,885	
1994–95	6,228	14,88,469	4,184	
1995–96	6,626	15,28,807	4,334	
1996–97	6,910	15,67,317	4,409	

Source: Compiled from various issues of *Tamil Nadu Electricity Board: A Glance*.

Figure 7.9
Energy use in agriculture

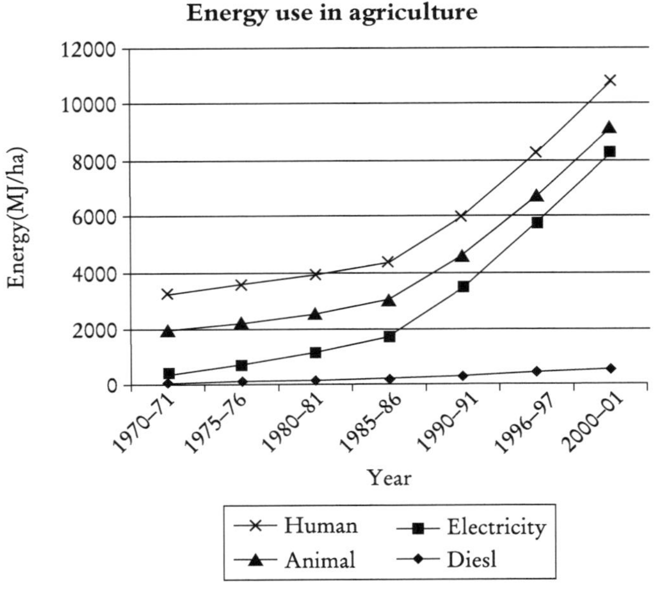

Source: Improvised from Kulkarni 2005.

table or low water supply from the wells that forces farmers to run pumpsets for longer hours. It strongly suggests that farmers tend to operate their motors even uneconomically due to the zero operating costs. Furthermore, because the owners of functioning wells are not paying the cost of power, they draw disproportionate benefits of power subsidies. On the basis of our survey of 38 well-owners in the Noyyal basin, we estimated that the 23 farmers who pumped water only once received a power subsidy of INR 7,110 while the 15 rich and successful well-owners who pumped the same amount of water twice received approximately INR 53,478. Existing power subsidies are thus heavily biased in favor of wealthy farmers.

Third, there is a high degree of polarization between water sellers and purchasers. In a separate study of the Vaigai basin, (Janakarajan 1997a), it was found that that a little more than three-fourths of the water purchasers are poor farmers whose landholding size is less than one hectare. A separate study in the Palar basin indicated that the extent of inequality in the distribution of land across all the cultivators (excluding landless population) is extraordinarily high, as reflected in a Gini coefficient of concentration of 0.88. Differences in Gini coefficients calculated separately for water purchasers and water sellers are, however, relatively small (0.34 and 0.40). This suggests that the "between-group" component of inequality (viz., between water sellers and water purchasers) is far greater than the "within-group" component (Janakarajan 1992). Furthermore, a vast majority of the water purchasers belong to the socially deprived castes; Scheduled Castes, the most deprived in the social hierarchy of India, constitute 27.3 percent of water purchasers (Janakarajan 1997a). This suggests that the agents involved in the water deals are sharply polarized socially as well as economically with unequal bargaining capacity.

The disparity between water sellers and purchasers often leads to subtle conflicts. Take, for example, the common informal rule that a water purchaser should purchase water only from the closest well-owner or, if all the well owners in a particular cluster agree, from the next nearest well-owner. This rule is intended to avoid conflicts, since increasing distance would require transporting water through the field channels of other farmers. It, however, places water purchasers at the mercy of adjacent landowners and conflicts often emerge when the rule is violated (Janakarajan 1992). The rule, in effect, places adjacent well-owners in a monopoly position. Furthermore, purchasing water from distant wells is difficult because the water purchaser needs to have equipment (such as a hose for transporting water) but has little or no guarantee that sellers will supply them regularly (Janakarajan 1997a). Second, unequal trading relationships in water markets often result in exploitation of the weaker agent. Water purchasers are often required to supply free or underpaid labor services to sellers. Refusal is impossible because well-owners can retaliate by stopping water supply in the middle of a cropping season (Janakarajan 1992). In several cases in the Vaigai basin, payments for water were made by way of rendering labor service and sharing a portion of the crop. While this can have advantages for cash-poor water purchasers and (in the case of crop-sharing) distributes the risk between water sellers and water purchasers, it is also open to abuse. Therefore, water markets become interlocked with the other labor, credit and product markets (Janakarajan 1992, 1997a). In addition, there are instances of water purchasers being forced to lease out parcels of their land to water sellers, at terms dictated by the latter. This is the case of *reverse tenancy* in which a lessee is seemingly more powerful than a lessor. Overall, although instances of open conflict between water purchasers and water sellers are infrequent, the former are often resentful of the latter. In addition, in some villages, water sellers collude in fixing the price of water high, which again generates conflict with water purchasers (Folke 1996; Janakarajan, 1992).

Agricultural water markets may now be declining. In the Noyyal river basin, sale of water for agriculture was never significant, but sale of water to industry is common. In the Palar basin, local agricultural water markets flourished until approximately a decade back. Since then, there appears to have been a significant drop in the extent of water sale (see Table 7.10). Farmers attribute this to progressive decline in groundwater table which has made it difficult for them to irrigate even their own crop, and to increase in groundwater pollution which has reduced all agricultural activities significantly. In other words, water sale has generally been a supplementary activity (the sale of excess supplies) with the primary goal of well-ownership being to fulfil one's own needs first. Table 7.10 presents information of surveyed wells reporting water sales in the Palar river basin. Roughly 10 percent of the sample wells in three villages reported water sales, supplying water to about 35 acres of land in an agricultural year. Only three of the

Table 7.10
Extent of water sale from the sample wells in the sample villages of Palar basin

Village	*Total No. of Sample Wells*	*No. of Sample Wells Reporting Water Sale*	*Gross Area to Which Water was Sold in 1998–99 (acres)*
Damal	49	7	20.5
Kathiavadi	41	1	2.5
Ramanaicken Pettai	43	5	11.7
Total	133	13	34.7

Source: Main survey (1998–2000).

eight villages in the Palar river basin had wells that reported significant water sales. In the remaining five villages, in all of which groundwater was found to be contaminated by tannery effluents, no water sale was reported.

The decline in agricultural water markets does not necessarily imply a decline in overall market activity in water. In the Noyyal river basin, there is a major water trade between rich well-owning farmers and urban industries (mainly dyeing and bleaching units). From two of the sample villages, viz., S. A. Palayam and Ugayanur, water is sold from 21 deep bores (with depths going up to 1,400 ft) to urban industries. According to our estimate, 100 million liters of water are easily transported daily from the villages around Tiruppur town. This can generate significant revenue: the rate per tanker with a capacity to hold 12,000 liters of water varies from INR 75 to INR 400, depending upon the season and the quality. Most of the wells were initially agricultural but have now been converted into commercial wells selling water. They belong to rich farmers who also irrigate a part of their landholdings. Some owners of industrial units also own their own deep borewells in this area. The effect of water sale on water users residing in the proximity of wells appears disastrous. Water yields from neighboring wells have declined significantly and members of well-owning farming households have almost stopped cultivation and are seeking jobs in the urban industries.

Overall, the evidence from Tamil Nadu indicates that local water markets are not a solution to growing water scarcity and, at least in the context of free power supply, have a limited impact on the incentive to reduce demand for water. At present, water markets seem to aggravate the problem of inequality and reinforce feudal relations.

A Return to the Larger Perspective

The detailed case study of groundwater issues in Tamil Nadu presented so far relates closely to the core issues facing groundwater development and management at the national level in India and Nepal.

That groundwater can play a critical role as a buffer against drought needs no elaboration. It is also now well established that crop yields in groundwater-irrigated areas are generally higher than those in areas irrigated by surface water sources, and that access to groundwater plays a critical role in agricultural development. In addition, strong arguments can be made in favor of the case that access to groundwater can play a major role in poverty alleviation and has done so in locations such as India (Moench 2001). Access to groundwater can reduce the risks of low crop yields and crop failures. By doing so, it can enable farmers (whether poor or wealthy) to begin a gradual process of agricultural intensification and prosperity that allows them to move out of poverty. The chief problem with groundwater, as this chapter details, is that access is not uniform. Even in areas where groundwater is close to the surface and major investments are not required to access it, groundwater development tends to parallel existing resource differentials within the rural society. Innovative farmers, i.e., farmers with exposure to new ideas and sufficient land to test them in their own plots (equivalent to "wealthy farmers"), tend to be the initial adopters new groundwater extraction technologies. As a result, the initial benefits from groundwater development tend to disproportionately go to those who are already economically better-off. This differential is exacerbated as the cost of accessing groundwater resources increases due to decline in groundwater levels, pollution or other factors. Early adopters have often accumulated sufficient resources to diversify their operations, to afford new equipments and to be able to deepen or drill wells as the water table declines. Later adopters and those whose overall resource base (land, education, access to capital, etc.) is limited, face major difficulties in maintaining their access. As a result, economic differentiation within rural communities increases. Differentiation and competition over scarce resources, in turn, increase conflict. The situation is particularly exacerbated by the fact that groundwater access is dependent primarily on an individual's socio-economic position. Unlike tank maintenance, maintenance of wells does not depend on community action. Furthermore, once an individual has access to groundwater, his incentive to contribute to community water supply systems is, for all practical purposes, nil. As a result, community water supply systems erode and so does the "safety net" for the poor in such systems as tanks, "spring" channels or large surface irrigation projects. In this context, groundwater, or rather the struggle to maintain access to it, can contribute to poverty and further socio-economic differentiation.

The groundwater situation in Tamil Nadu is affected by the hard-rock nature of the geology. Because wells are dug into hard rocks where storage is low and water yields from wells depend heavily on chance (whether or not wells hit the productive fracture zones), the dynamics of groundwater access are different from areas underlain by alluvial aquifers. Several factors contribute to this difference:

(*a*) *High, location-dependent risk*: The risk of investing in unproductive wells is far higher in hard-rock areas than in alluvial areas. In most alluvial areas, regional groundwater levels are the primary factor determining the ability to access groundwater: one just needs to drill wells to sufficient depths. In hard-rock areas, however, fracture patterns are often highly variable. As a result, the chance of success in tapping a productive zone depends on the financial resources to drill multiple bores *and* on a large landholding with multiple locations where it may be possible to drill a well.

(*b*) *Low storage*: The low storage in hard-rock areas heavily biases benefits toward early adopters. Because storage is low, decline in water levels occurs rapidly and those who dig or drill the first wells are far more likely to obtain water at a reasonable cost than those who attempt to do so later.

(*c*) *Low well yields*: Because well yields tend to be low in hard-rock areas, little surplus is generally available beyond the amount needed to irrigate the lands immediately adjacent to wells.

The problems of low well yields, low storage and the high locational risk particularly associated with hard-rock aquifers have important implications for the nature of water markets. Many of the studies on water markets in India have been done in the deep alluvial aquifers of Gujarat. There, although water levels are falling, the capacity to pump water from any given well tends to be relatively high and relatively uniform within a given area. As a result, small, medium and even large farmers are often able to reliably pump significantly more water than they can use for irrigating their own lands. There is, as a result, often a strong incentive to sell excess water. Since power is charged at a flat annual rate based on pump horsepower, there is no marginal cost, and sale of any excess water at any rate reduces average costs. In many such locations, the bargaining position of both buyers and sellers is relatively equal. This type of dynamic can lead to incentives for water sales at rates below the full cost of well development (Moench 1996, 1995; Shah 1993). The situation is fundamentally different in hard-rock areas where well yields are low and often vary greatly across seasons. In this situation, surpluses are far smaller and tend to vary greatly across seasons and locations. It is a seller's market in which the bargaining position of water buyers is weak. This is probably a major factor underlying the interlocking of other agricultural markets with those for water.

Where does this leave us with respect to global and local debates over the role of groundwater markets? This role is discussed in detail in Moench and Janakarajan (2006), which deals with markets and commodity chains. It is, however, important to emphasize the observation from the fieldwork that water markets in Tamil Nadu and in the rest of India are self-initiating institutions. They were not created by government interventions and their characteristics are difficult to influence through government policies. They *exist* as informal institutions outside the formal legal or regulatory frameworks of the government. In addition, their characteristics vary greatly across regions and locations. Furthermore, as availability of groundwater changes, the characteristics of water markets, too, change. As a result, while it is important to understand the impact of groundwater markets on access to a key resource for local populations, there is probably not much that can be done to influence their dynamics under existing circumstances.

This observation on groundwater markets raises an important question: how is civil society going to respond to escalating and competing demands for shrinking groundwater resource base?

Tushaar Shah illustrates the issue at a national scale in a diagram he prepared for the book *Groundwater and Society* (Burke and Moench 2000: 66). This diagram illustrates the transitional nature of groundwater development and use across India. Initially, groundwater development catalyzes change and the development of an intensive agricultural economy. Then, as development levels reach or exceed the sustainable levels, the economy that has grown on intensive groundwater irrigation must transition. In some areas, intensive agriculture based on groundwater use may be sustainable. In other areas, limitations on the physical availability of groundwater will force a transition. How this transition occurs is, perhaps, the biggest question facing groundwater management. Will it be possible for rural populations to make a planned (or at least smooth) transition to other forms of economic activity and limit groundwater extraction to sustainable levels, or will the transition be driven by the types of dynamics currently seen in the case of Tamil Nadu? In the Noyyal basin, small well-owning farmers in Orathapalayam and Karaipudur have been so badly affected by groundwater pollution that they are being forced out of agriculture to seek jobs in the urban areas. A similar dynamic is occurring in Ugayanur and S. A. Palayam villages where farmers have been driven out of agriculture due to their inability to stay in race of competitive well-deepening. In S. A. Palayam, 16 out of the 54 wells surveyed have gone dry and are not in use. Their owners have lost in the race of competitive well-deepening and are heavily indebted. Over 60 percent of well-owners in Tamil Nadu are small and marginal farmers with landholdings of less than 5 acres. Their economic survival is threatened by groundwater pollution and overdraft. How rural populations of this type can transition to more sustainable livelihoods has emerged as a critical question throughout much of India.

Power Subsidies and Groundwater Pollution

As documented earlier, benefits from the existing power subsidies are mostly enjoyed by wealthy farmers. In addition, the provision of free power encourages highly inefficient water use practices and groundwater overdraft. This is a clear case where policy reform is required. Reform must, however, also address the issue of transition. At present, even if INR 0.5 is charged for every unit of electricity consumption, many small farmers will have to close down their wells because of uneconomical conditions (Janakarajan forthcoming). Continuing subsidies that primarily benefit the wealthy farmers and encourage unsustainable patterns of groundwater use would be counterproductive, but the displacement caused by policy reform must also be recognized and addressed.

Pollution is also a critical issue. This will not be addressed through policy reform alone. Though existing pollution laws in Tamil Nadu, as in the rest of India, are sufficient to enable corrective action, they are not, however, generally enforced. As has been argued elsewhere (Moench et al. 1999), social auditors, the independent voices that expose uncomfortable truths in society, need to build pressure on governments and others to act.

≈

Acknowledgements

This essay is the outcome of a bigger study, "Local Water Supply Options and Conservation Responses," carried out with the financial support of the International Development Research Center (IDRC), Canada. We gratefully acknowledge the research assistance provided by K. Sivasubramanian and G. Jothi. This essay was earlier published in 2006 under the same title in *Economic and Political Weekly*, 41(37): 3977–87.

References

Amarasinghe, U. A., L. Mutuwatta and R. Sakthivadivel. 1999. *Water Scarcity Variations within a Country: A Case Study of Sri Lanka*. Colombo: International Water Management Institute (IWMI).

Bhatia, Bela. 1992. "Lush Fields and Parched Throats: Political Economy of Groundwater in Gujarat," *Economic and Political Weekly*, 27(51&52): 19–26.

Burke, J. and M. Moench. 2000. *Groundwater and Society: Resources, Tensions, Opportunities*. New York: United Nations (UN).

Central Ground Water Board (CGWB). 1995. *Groundwater Resources of India*. Faridabad: CGWB, Ministry of Water Resources (MoWR), Government of India (GoI).

Dhawan, B. D. 1995. *Groundwater Depletion, Land Degradation and Irrigated Agriculture in India.* New Delhi: Commonwealth Publishers.

Folke, Steen. 1996. "Cauvery Delta: Canal and Well Irrigation and Conflicts over Land and Water." Unpublished paper.

Food and Agriculture Organization (FAO). 2000. *Agriculture: Towards 2015/30.* Interim report. Rome: Global Perspective Studies Unit, FAO. Also available online at http://www.fao.org/fileadmin/user_upload/esag/docs/Interim_report_AT2050web.pdf (accessed September 22, 2013).

Government of India (GoI). n.d(a). *Ninth Five year Plan 1997–2002,* vol. 2: *Thematic Issues and Sectoral Programmes.* New Delhi: Planning Commission, GoI.

———. n.d(b). *Indian Agricultural Statistics, 1985–86–1989–90,* vol. 1. New Delhi: Ministry of Agriculture, GoI.

Government of India. 1992. *Report of the Committee on Pricing of Irrigation Water.* New Delhi: Planning Commission, Government of India.

Government of Tamil Nadu (GoTN). 1998. *Season and Crop Report for the Year 1997–98.* Chennai: GoTN.

Harriss, John. 1982. *Capitalism and Peasant Farming: Agrarian Structure and and Ideology in Northern Tamilnadu.* Bombay: Oxford University Press.

Janakarajan, S. forthcoming. "The Politics of Power Subsidisation: The State of Power Sector in Tamil Nadu," *Water Nepal,* 8(1/2): 39–62.

———. 2000. "Issues Relating to Official Data Base on Groundwater Resource in Tamilnadu," *Review of Development Change,* 5(2): 385–407.

———. 1999. "Conflicts over the Invisible Resource in Tamilnadu: Is There a Way Out?," in M. Moench, Elizabeth Caspari and Ajaya Dixit (eds), *Rethinking the Mosaic: Investigations into Local Water Management,* pp. 123–60. Kathmandu: NWCF; and Boulder, CO: ISET.

———. 1997a. *Conditions and Characteristics of Groundwater Irrigation: A Study of Vaigai basin in Tamilnadu.* New Delhi and Chennai: Planning Commission, GoI, and Madras Insitute of Development Studies (MIDS).

———. 1997b. "Consequences of Aquifer Overexploitation: Prosperity and Deprivation," *Review of Development and Change,* 2(1): 52–71.

———. 1993. "In Search of Tanks," *Economic and Political Weekly,* 28(6): A53–61.

———. 1992. "Interlinked Transactions and the Market for Water in the Agrarian Economy of a Tamil Nadu Village," in S. Subramanian (ed.), *Themes in Development Economics: Essays in Honour of Malcolm Adiseshiah,* pp. 151–201. New Delhi: Oxford University Press.

———. 1986. "Aspects of Market Inter-relationships in a Changing Agrarian Economy." PhD thesis, University of Madras, Chennai.

Kulkarni, S. D. 2005. *Country Paper-India Agricultural Mechanisation-Present Scenario and Perspective.* Bhopal: Central Institute of Agricultural Engineering.

Ladha, J. K., K. S. Fischer, M. Hossain, P. R. Hobbs and B. Hardy (eds). 2000. *Improving the Productivity and Sustainability of Rice-Wheat Systems of the Indo-Gangetic Plains: A Synthesis of NARS-IRRI Partnership Research.* Manila: International Rice Research Institute (IRRI).

Lindberg, Staffan. 1999. "While the Wells Ran Dry: Tragedy of Collective Action among Farmers in South India," in Stig Toft (ed.), *State, Society and the Environment in South Asia*, pp. 266–296. Madsen and London: Curzon Press.

Malik, R. P. S. 1993. "Electricity Pricing and Sustainable Use of Groundwater: Evaluation of Some Alternatives for North-West Indian Agriculture." Paper presented at the workshop, "Water Management: India's Groundwater Challenge," Vikram Sarabhai Centre for Development Interaction (VIKAST), Ahmedabad, India, December 1 and 16.

Moench, M. (ed.). 1995. *Electricity Prices: A Tool for Groundwater Management in India.* Ahmedabad: VIKSAT.

Moench, M. 2002. "Water and the Potential for Social Instability: Livelihoods, Migration and the Building of Society," *Natural Resources Forum*, 26(3): 195–204.

———. 2001. "*Groundwater and Poverty: Exploring the Links.*" Paper presented at the workshop, "Intensively Exploited Aquifers," Royal Academy of Sciences, Madrid, Spain, December 13–15.

———. 1996. *Groundwater Policy: Issues and Alternatives in India.* Colombo: International Irrigation Management Institute (IIMI).

———. 1994. "Hydrology under Central Planning: Groundwater in India," *Water Nepal*, 4(1): 98–112.

———. 1993. "Electricity Pricing and Groundwater Use Efficiency: Critical Linkages for Research," *Pacific and Asian Journal of Energy*, 3(1): 121–29.

———. 1992a. "Chasing the Watertable," *Economic and Political Weekly*, 26: A171–77.

———. 1992b. "Drawing down the Buffer," *Economic and Political Weekly*, 17(13): A7–14.

Moench, M. and S. Janakarajan. 2006. "Water Markets, Commodity Chains and the Value of Water," *Water Nepal*, 12(½): 81–114

Moench, M., E. Caspari and Ajaya Dixit (eds). 1999. *Rethinking the Mosaic: Investigations into Local Water Management.* Kathmandu: Nepal Water Conservation Foundation and Institute for Social and Environmental Transition.

Nanjamma, C. B. 1977. "Adoption of the New Technology in North Arcot District," in B. H. Farmer (ed.), *Green Revolution? Technology and Change in Rice Growing Areas of Tamilnadu and Sri Lanka*, pp. 92–123. New York: Macmillan.

Palanisamy, K., R. Balasubramanian and Mohamed Ali. 1996. *Management of Tank Irrigation by Government and Community: An Economic Analysis of Tank Performance in Tamilnadu.* Coimbatore: Water Technology Center, Tamil Nadu Agricultural University (TNAU).

Rao, D. S. K. 1993. "Groundwater Over-exploitation through Bore Holes Technology," *Economic and Political Weekly*, 28(52): A129–A134.
Rosegrant, M. W. and C. Ringler. 1999. *Impact of Food Security and Rural Development of Reallocating Water from Agriculture*. Washington DC: International Food Policy Research Institute (IFPRI).
Shah, T. 1993. *Groundwater Markets and Irrigation Development: Political Economy and Practical Policy*. Bombay: Oxford University Press.
Singh, C. 1991. *Water Rights and Principles of Water Resources Management*. Bombay: Indian Law Institute and N. M. Tripathi Pvt Ltd.
———. 1990. *Water Rights in India*. New Delhi: Indian Law Institute.
Vaidyanathan, A. and K. Sivasubramanian. 2004. "Efficiency of Water Use in Indian Agriculture," *Economic and Political Weekly*, 39(27): 2989–96.
Vaidyanathan, A. 2001. *Tanks of South India*. New Delhi: Centre for Science and Environment.
———. 1996. "Depletion of Groundwater: Some Issues," *Indian Journal of Agricultural Economics*, 51(1&2): 184–92.
Vaidyanathan, A. and S. Janakarajan. 1989. *Management of Irrigation and Its Effect on Productivity under Different Environmental and Technical Conditions*. New Delhi and Chennai: Planning Commission, GoI, and MIDS.
World Bank, The and GoI. 1998. *India – Water Resources Management Sector Review, Groundwater Regulation and Management Report*. Washington DC and New Delhi: The World Bank and MoWR, GoI.

PART IV

Water Conflicts and Cooperation in South Asia

8

Water Conflicts, Contending Water Uses and Agenda for a New Policy, Legal and Institutional Framework*

K. J. Joy and *Suhas Paranjape*

> The unfolding scenario for water use in many parts of the world is one of increasing concern about access, equity and the response to growing needs. This affects relations between rural and urban populations; upstream and downstream interests; agricultural, industrial and domestic sectors; and human needs and the requirements of a healthy environment (World Commission on Dams 2000).

This statement from the report of the World Commission on Dams succinctly sums up the increasing conflicts over water. Though the "water wars," mentioned in a chance remark by the then UN Secretary General Kofi Annan have not materialized,[1] water is radically altering and affecting political boundaries all over the world, between as well as within countries. Indications are that water conflicts of different types are going to get worse and they pose a significant threat to economic growth, social stability, security and ecosystem health (Joy et al. 2007).

This essay primarily deals with one type of water conflict, namely, contending water uses and users. The essay is organized

*The essay draws upon three previous articles written by the authors: Joy et al. 2007; Joy and Paranjape 2009, 2007.

into five sections. Section One discusses the immediate context of the status of water resources in India. Section Two situates contentions over water use within an overall typology of water conflicts. Section Three argues for an integrated approach to contending water uses and attempts to outline a normative framework to deal with issues like equity, integration of local and exogenous water, surface and groundwater, variable and assured water, water rights, allocations, and transfers and related issues. Section Four discusses some of the critical issues related to the legal and regulatory frameworks with regard to water, rights, regulation and globalization, and the interconnectedness of law and policy in the water sector.

The Context: State of Water Resources in the Region

Global assessments of water availability indicate that the per capita water availability would decrease and the overall demand for water in India would increase keeping up with the increasing population as well as the increased demand from certain sectors. For example, a study by International Water Management Institute (IWMI) has tried to project the scenario of water availability in the year 2025 and has divided the countries into categories according to the relative availability of water.

The first category includes those countries that are most water scarce and in 2025 will not have enough water to maintain 1990 levels of per capita food production from irrigated agriculture and meet industry, household, and environmental needs. The study notes that while India and some of the other South Asian countries will not have major water problems on average, there will be massive regional variations in water availability. Though India is placed in the category of countries that will have to increase water development modestly overall – on an average, by only 5 percent – to keep up with 2025 demands, a sizable portion of its population (280 million people in India in 1990) is placed in the category of absolute water scarcity (Seckler et al. 1998).

Of course, not all problems related to water can be reduced only to "scarcity"; the issue of distribution is also equally important. However, decreasing availability of water is also real and most of the conflicts over water involve a perceived unfairness in the distribution of this "scarce" resource. The extent of the crisis in the water sector in India is evident from the dwindling water availability and the increasing conflict around it. The country's utilizable freshwater resource is estimated at about 110–12 million hectare-meters (ha-m). Of this, little less than half (53 million ha-m) is currently utilized. The renewable water resource in terms of annual rainfall, estimated on the basis of average precipitation, comes to 400 million ha-m. Of this, 185 million ha-m is available as surface storage, 50 million ha-m is stored underground (groundwater), and 165 million ha-m is stored in the soil as root-zone soil moisture (Paranjape and Joy 2008). One-third of the country, about 109 million hectares (mha), mostly in the central and western parts, is categorized as drought prone and faces a serious drinking water crisis.

Contending Water Uses within a Broad Typology of Water Conflicts

Some themes of water conflicts like contending water uses, equitable access and allocation, water quality and pollution, and dams and displacement can be found anywhere in the world, but are especially important in the developing countries.[2] Recently, privatization of water in various forms has become another arena of conflict and contestation in developing countries, as there is a growing rethinking about the role of state in providing water. Developing a typology is not an easy task, as many conflicts fit into more than one theme. Since water conflicts are often a multifaceted microcosm of wider conflicts, it is rather difficult to identify any one aspect as dominant and thus it is not possible to make the themes mutually exclusive. However, as a working hypothesis we may locate the conflicts surrounding contending water uses within the following major types of conflicts in the region.

Contending Water Uses

Contending water uses or contention between *different* kinds of uses is an important, emerging type of water conflict in India. Water, as noted earlier, is a common pool resource; hence, when the same unit of water is demanded for different kinds of uses, we have a contestation and a potential conflict. In the Keoladeo National Park in the state of Rajasthan, a classic case of contested water use,[3] the contestation is between the needs of the Bharatpur wetlands, a spectacular bird sanctuary which is a World Heritage and Ramsar site, and the irrigation needs of the local farmers; the conflict continues to be resolved every year, in an ad hoc manner. There has been an increasing awareness of the need to keep minimum environmental flows within the river systems and this has become an issue of contestation between the environmentalists and water resource engineers. Similarly, the increasing demands from industry and urban areas are putting pressure on the available water resources (and with increasing population the per capita availability of water is showing a drastic downward trend) in most parts of India.

Equity and Access

Another related theme is equity which focuses on issues of distribution and access of different users but *for the same kind of use*, unlike the first theme that deals with *different* contending uses (Paranjape and Joy 2007a).[4] Equitable access to drinking water or irrigation water is a case in point and covers a wide variety of issues including contestation over and between old and new water rights, old and new projects, tail-enders and head-reachers in the designed irrigation commands, *dalit*s and upper castes and different ethnic groups, etc. In southern Maharashtra, a western Indian state, drought-affected farmers under the banner of Shetmajoor Kashtakari Shetkari Sanghattana (Organization of the Agricultural labourers and Toiling Peasants) have been organizing various types of agitations to restructure the newly initiated irrigation projects in the area on more equitable lines.[5] Increasingly, the different sections of the society are demanding their share of water and in countries like India the contestation over equitable access to water has taken a much more organized form.

Almost 80 years ago, Dr. Bhimrao Ambedkar launched a water satyagraha in Mahad, Maharashtra, by marching to the Chavadar Tank to open up all public watering places to *dalit*s.[6] The conflicts in Mangaon and surrounding villages, very close to Mahad, five to six years ago, show how in a drought year centuries of caste-based oppression and prejudice, deep-rooted cultures and traditions, rear their heads once again to deny water to the *dalits*.[7]

Pollution-induced Conflicts

Water quality or pollution is fast emerging as another arena of conflict. Earlier, these issues were treated as inevitable consequences of growth and industrial development and, therefore, largely ignored. Also, non-point source pollution due to increased use of pesticides and fertilizers is also posing a major threat to the ecosystem and the livelihoods of those dependent on these water sources. However, growing scale, increased awareness and active civil society engagement has brought water quality conflicts more and more to the forefront. The main issue here is how and in what form do users return water to the ecosystem. Polluted water returned by users causes problems to downstream users, and decreased freshwater availability causes economic loss, social distress and ill health. The Noyyal[8] and Palar[9] basins in Tamil Nadu, India, are classic cases of industry-induced pollution. Water pollution impacts both ecosystems and peoples' lives and livelihoods, so much so that any possible solution to conflicts related to water quality needs to address both ecosystem needs and livelihood needs.[10]

Dams and Displacement

Conflicts over dams and displacements are relatively well publicized and better documented. For example, the case of Sardar Sarovar Project in India[11] has become the rallying point for all those who are against large dams globally. Struggles against dams have posed a basic question: whether large dams are required or not. Quite often, response to this question has led to polarized positions – large versus small dams, or pro-dam versus anti-dam. Most of the South Asian countries have a short, well-defined monsoon in which most of the rainfall occurs. Since the water received from rainfall in this short season has to meet the water needs for the rest of the period, probably a certain amount of storage through dams to meet the water requirements during non-monsoon months becomes a necessity. In the drought-prone regions, exogenous water from large and medium dams may be needed to supplement and strengthen local water harvesting and that their integration is the way to avoid dividing the poor and pitting them against each other as drought-affected beneficiaries versus displaced victims.[12] The issue is whether we can utilize water from larger sources without creating disruptive submergence and displacement, and whether we can rehabilitate the displaced people as part of an area development plan.[13] There are also examples, may be very few, showing that with their combined strength and organization, and under a creative leadership, project-affected people probably can come up with innovative options for their own development-oriented rehabilitation. An interesting example of this effort is the case of Tawa Fishing Cooperative Society in Madhya Pradesh.[14]

Water Privatization

Privatization of water is an important upcoming arena of conflict in many countries in Asia, Latin America and Africa. The present discourse on water privatization calls for a nuanced understanding of the issue – source privatization and privatization of service delivery cannot be equated – and also brings to the fore the underlying issues of equity and assurance that are most threatened by privatization or what goes on under the rubric of public–private partnerships (PPPs), the latest buzzword in water management. The current debate on water privatization is highly polarized between two well-entrenched positions in favor and disfavor of privatization, and there seems to be very little attempt to explore the middle ground of seeing water as both a social and economic good.[15] The famous Plachimada case[16] in the southern state of Kerala, where the Coca Cola bottling plant obtained permission to use groundwater despite the opposition from the local Gram Panchayat, and the case of Sheonath river[17] in the central Indian state of Chhattisgarh are cases of conflict around privatization of water in India. They also raise the questions about whether or not privatization of rights and entitlements takes place under the garb of privatizing services.

Transboundary Water Conflicts: A Major Cause of Regional Tension

Conflicts between countries are generally classified as transboundary conflicts. However, in India, water is a state subject and states enjoy considerable autonomy in deciding on issues related to water. In India, almost all inter-state rivers have become arenas of contestation and conflicts amongst the riparian states. The conflict between Karnataka and Tamil Nadu over the sharing of the Cauvery waters is the most intense of all. The recent verdict of the Cauvery Tribunal apportioning the river water amongst the riparian states has only added further fuel to the issue. However, there are transboundary water conflicts over sharing international river waters as well. The India–Pakistan dispute sparked by the Baghlighar project can be traced to the Indus Water Treaty (IWT) signed between them in 1960 (Sinha 2007). Another important case of transboundary contention – sharing of Ganga waters – is that of inter-basin diversions of the Himalayan rivers and their implications for Indo-Bangladesh relations (Sen 2007).

Within these broad types of water conflicts, the theme of contending water uses and users covers the first two types of conflicts. The first is the contention between *different* kinds of uses and range from drinking/domestic versus agriculture and agriculture versus industry to urban versus rural, ecosystem use versus consumptive use, etc. The second type is the conflict between different users but *within the same kind of use*, for example in agriculture. Contestation over and between old and new water rights, old and new projects, tail-enders and head-reachers in a command area, and between *dalits* and upper castes or different ethnic groups are all examples of this type.

Need for an Integrated Approach

All these conflicts can be seen as involving the issue of what "ought" to be a "fair" allocation between or within different uses. And none of them can be discussed, let alone resolved, without some discussion of the principles, or the normative framework, that may, explicitly or implicitly, be the basis on which these allocations can be made. The first important requirement of the framework is to treat equity as central and as a starting point. Equity cannot be added on as an afterthought. In the suggested framework, equitable access to water is treated as a matter of *minimum* assurance to all of the water required for livelihood needs, *irrespective* of their ownership of assets. Minimum water assurance is seen in this framework as a *right* that vests in *people* by virtue of their right to an adequate livelihood, and not by virtue of their ownership of land or other assets, as is the case with conventional approaches towards a right to water. There is some basis for this principle in the traditional practices based on natural equity. Such practices may provide minimum water assurance to all farmer households. However, it may not always be easy to extend these practices to include minimum water assurance to the landless or women. Special efforts will be needed to bring such disadvantaged groups within the ambit of minimum water assurance and thereby give them the right of equitable water access.

The degree and extent to which this right can be given has to be assessed within a framework of assurance of needs. Rights cannot be truly dissociated from need; especially, in talking about rights of access in terms of entitlements or endowments, we cannot dissociate them from some definition of need distinct from want. While this demarcation will always remain value laden and contested, it is not one that can be avoided. We use here a definition of need that includes basic livelihood needs in terms of biomass. This approach has been elaborated elsewhere;[18] its main advantage is that it provides us with what should count as a basic service necessary to fulfil basic livelihood needs. This helps define a minimum water assurance that must be provided for at a reasonable cost and dependability, and the rest of the portion, after adjustment for ecosystem needs, only may then be treated as water available for allocation and for other economic uses. *In a sense it separates the entitlement into one based on rights and one based on economic opportunities.* This separation is an important separation because water has features of both basic right and economic good, and law must first distinguish between and demarcate them.

Equitable access also has associated costs and requirements, and a commitment to equity necessarily implies meeting those requirements and costs. Changing social arrangements to suit equitable access is the most cost-effective measure. An illustrative example is that of the old *phad* system prevalent in the eastern part of Maharashtra, wherein every landowning family in a village was assured land in the *phad*, a demarcated portion of the village land that could be irrigated by gravity.[19] If such arrangements do not work out, it may be necessary to carry water over longer distances, to provide for lifts and pumping energy for a portion of the service area. That does make it important to minimize such costs (economic, social as well as energy costs), socially as well as through system planning and technology.

The lack of integration of various kinds of water resources is at the root of the problem in extending water service equitably. Conventional thinking has often actively discouraged such integration for the ease of management, but new social movements and approaches, too, have tended to reject such integration and, indeed, exogenous water, as well as large sources and dams altogether. However, in our opinion, integration would be a must for both if they were to synergize the exclusive advantages that each source of water has to offer and the other does not have. There is, however, a broad principle involved in the integration of these two sources. It implies that the local water source is developed to its fullest potential, the local system is built around it and the exogenous water provides supplements to *the system* as a whole rather than individual irrigators. Watershed development can become an integral part of the whole process of large source development that is built around it. It is a pity that this aspect is not receiving the attention it should.[20]

Similarly, there is a need for an integrated approach to groundwater and surface water management. The separation of groundwater and surface water management has been an explicit assumption and a *desideratum* of the conventional approach to surface irrigation systems. However well intentioned and desirable it may have been, their separation, in actual control as well as ownership and utilization rights, has had a pernicious effect. Their integration in Ozar (Nashik district, Maharashtra) has demonstrated, to some extent, that it can lead to much tighter control over the timing and quantum of water application and help service a larger area much more effectively.[21]

Water Rights, Allocations and Transfers

Water rights, allocations and transfers depend on our approach to water as property. It has been argued that water should be seen as a social, common property as distinct from state property. The state acts as the trustee of that property. It is, therefore, the duty of the state, as a trustee, to ensure access to a fair share of that water for reasonable use to every citizen as a right. An integrated approach suggests that citizens have the right to water use (and not ownership over water) subject to the concepts of fair share and reasonable use. The main problem lies in determining water allocations, what a fair share is and what constitutes reasonable use. There are two aspects to this: the norms that govern the determination of needs that we have discussed earlier and the process through which they shall be determined.

In respect of the norms that determine regional water allocation, we should treat water allocation in proportion to the needs of the population in the region as the starting point from which to evolve a principle of a fair share of water resources available for use. We also need to take into account the priorities of water use. The order of priority that should normally be observed is water for drinking (including water for cattle and domestic use), for environmental needs and for livelihood needs, as also a flexible order of priority between extra agricultural production (over and above livelihood needs), industry, tourism and other uses according to the particular character of a region under consideration.

Legal and Regulatory Frameworks

Indian jurisprudence and case law on water has evolved in many different directions and from different underlying conceptual frameworks.[22] Not all of them are compatible with each other and may even lead to contradictory judgments. One major line of thinking treats water as property, most often to be enjoyed by virtue of land ownership. This is the line of thinking that also bestows on a land-owner the right to extract as much groundwater as he/she may from any well within his/her land. It tends to see water as classical property. Even here, there are two strands of thought that may at times be in conflict. While one stream tends to treat water, especially groundwater, as property tied to land, the other strand of thought treats it as part of what has come to be called as the "eminent domain"[23] which asserts state supremacy and control of water resources that amounts *de facto*, if not *de jure*, to state ownership over water.

Another strand of thinking, increasingly being deployed after Independence, sees water from within a framework of human, natural or basic rights. Within the rights perspective, water seems to be most frequently treated as part of the right to life and *ipso facto* a basic right. There is no *explicit* recognition of a right to water in the Constitution of India, though it can be treated as implicit in many ways (Iyer 2009). The diversity of views on water and the law is also further amplified by the division of power that the Constitution specifies. There are few clauses of the Constitution that deal with water. Water falls in the State list and for all practical purposes the Union has very little role to play, except in the case of disputes over the waters of inter-state rivers, though it could be argued that Entry 56 is worded in such a manner that it gives the Union considerable leeway in regulating inter-state matters provided it is backed by legislation in the Parliament.[24]

Thus, there is a lack of a comprehensive legal framework based on common principles. Each state has enacted its own laws on water, and we may add to this the fact that at the lowest level, especially in rural and tribal areas, water-related practices are often governed by religious and customary rules.[25] An important reason for the diversity (or legal pluralism) is the immense diversity of water regimes in India, which requires a corresponding diversity in social and political arrangements around water. However, we still need an underlying set of common principles and a common normative framework underlying water sector policy and legislation in the country. Nowhere does this matter as much as in deciding on contending water uses.

Changing Legal Terrain: Rights, Regulation and Globalization

In recent times, the legal framework generally, and especially around water, is changing under increasing pressure from two directions: one from the rights perspective and the people's movements from below; and the second from above due to the reform process initiated in the early nineties by the world-wide process of globalization. There are at least three rights issues in relation to water besides the one that comprises the right to life and involves drinking water and water for sanitation. First is the issue of livelihood and the water that is required to earn a livelihood. A farmer requires water for his farm, and a tanner requires water for cleaning and curing the hides. Almost every kind of livelihood activity requires some water in one form or the other and this raises the issue of whether the right to that water should not be treated as part of the right to life, since earning a livelihood is essential to living? Whether or not water for livelihood is treated as a basic right, it certainly needs to be included as a priority use.

Moreover, there is now growing pressure from below to treat equitable access to water as part of equity in assuring livelihoods for all. The pioneering Pani Panchayat movement and the South Maharashtra movement have made it one of the major planks of their campaigns.

The second is the increasingly recognized need for environmental flows, which is seen in some ways as a right of ecosystems. The issue involved here is whether or not we need to ensure that some flows remain unbound and "unused" so that ecosystem services may be regenerated. However, those with environmental leanings tend to prioritize preservation of ecosystem integrity over simple regeneration of ecosystem services. The issue remains unresolved, but it is becoming increasingly clear that we can no longer treat unbound flows as "wasted" because it "flows out to the sea" and must arrive at a consensus on the level of flows that may be treated as necessary for environmental needs.

The third issue is the "right" to use natural resources as economic resources for entrepreneurial gain. This right is also often seen as important in itself. Indeed, there is increasing pressure to recognize natural resources, including water, as an economic good and remove all restrictions that supposedly "distort" its operation as an economic good. Here, too, there are very different approaches in play. On the one hand, the globalizers want water to become a freely priced economic good and believe that doing so will actually allow for more efficient allocation and thereby better fulfillment of all rights. On the other hand, there is an equally large number of people, including those from the developed countries, who would argue for the operation of economic laws in respect of water in a restricted and regulated space.

Much of the present discourse on a right to water stops at the level of treating it as part of a right to life and does not go on to consider other dimensions of the right to water. We would suggest that the legal framework must take as its starting point an articulated hierarchy of these rights. Thus, we have at the bottom the right to water as part of the right to life (water of adequate quality and quantity for drinking, other domestic purposes, sanitation and livestock) as an *absolute* right so that its fulfilment forms the precondition for any other water use or right. This should be followed by the water needed for the fulfilment of environmental needs and livelihood needs which together form a set of relative rights. It may be conceded that there are greater options and greater scope for variation and flexibility than in the right to water as part of the right to life. Thus, this tier needs to be taken together, i.e., in case of shortage, *both* uses share the brunt of the shortage and, to some extent, may even share it with the following uses. That is why, we class it as a relative right rather than as an absolute right. Nevertheless, this tier must be regarded as having a higher priority than the following one, so that in case of shortages the following tier shares a much larger share of the brunt.

Since the 1990s, there have been increasing international pressure directed towards the so-called reform process, and the legal frameworks in the country have been changing in response to that pressure as well. There has been a general drive towards a managerial change, a change in favor of users' participation including user contribution to capital costs, a change in pricing policy oriented towards removal of subsidies and a move away from state control to a regulatory regime overseen by independent regulators. The World Bank and other donors have been making some of these measures a requirement for disbursing their funds, and this has acted as an important driver in initiating this change.

The Maharashtra Management of Irrigation Systems by Farmers (MMISF) Act passed in 2005 provides for statutory formation of Water Users' Associations (WUAs) in all the command areas served by canal irrigation. Under the Act, the state and the WUAs enter into a Memorandum of Understanding (MoU) that fixes a volumetric quota for the WUAs, and the actual quota for any year is determined on the basis of reservoir filling and availability of water (GoM 2005a). What is important to note is that the MoU provides for a definite binding and takes us closer to a mutual arrangement than a simple unilateral quota granted at the state's discretion. It is also important that the water charges are related to the quantum of the water delivered to the WUAs rather than to a localized crop pattern and area-based charge. The latter arrangement almost invariably meant a flouting of the designed crop pattern. The WUAs are thus free to plan their crop patterns on the basis of the volumetric supply they receive, and the need for spurious double accounting of water, which created many avenues of corruption, is now minimized.

However, there are a number of problems with the Act insofar as we see it as an instrument for restructuring the water sector along more sustainable and equitable lines. Though we can hardly expect equity to be factored into water sector legislation by the present state, it does not provide for enough flexibility for water users to change the basis of water distribution from the proportionality of land owned within the command area to a more equitable concept based on minimum water assurance; and to change the basis of those eligible for receiving water and extend water service to the needy people outside the present designated command areas who could also be served if the distribution basis becomes more equitable. The Act in its present form, thus, would simply result in a freezing of presently iniquitous allocations and relations within canal-irrigated areas.

The first problem of the Act is that in spite of its progressive features, it is based on a static and classical concept of property.[26] As we have discussed earlier, there are many ways in which water differs from classical static natural resources and needs a new type of concept of dynamic entitlements related to the type of water use.

In some ways, this parallels what Iyer (2009) has identified as use rights rather than property rights. This switch is a welcome one, but it must also make space for the dynamic re-allocation, which is required by a fast changing water use pattern that is characteristic of modern societies. For this, allocations must be provisional, valid for a certain period, and there must be scope and mechanisms for re-allocation through periodic re-negotiation. This is even more important if we are to incorporate the issue of contending water uses. Contending water uses need to be recognized as a problem that cannot be wished away but must be tackled through a process of negotiation. There is scope for neither in the legal enactments that are taking place.

From Community Management to Private Property Rights to Participative Allocation and Governance

Traditionally, water has been treated as a community resource and has been subject to community norms and rules that govern its access and use. It has never been treated as private property, i.e., as property with untrammeled rights of use and misuse if the owner so wishes. It has never been treated as merely a resource but has taken on many more roles. Civilizations have grown around it, searched for its sources and nurtured it, and there is evidence that those that have failed in nurturing it have also failed as civilizations. It has been as much worshipped as utilized and the worship has always tempered its use. The advent of modernity has dissolved this community nexus that regulated affairs between people at various levels and has left the abstract individual as property holder standing in opposition to a codified state, with, in theory, no community mediating this relationship. The concept of freedom includes the untrammeled right to do with one's property as one pleases. This framework simply does not suffice for water as a resource.

Nevertheless, this is the framework that the British brought to bear on land and water in India, and through the system of colonial courts and laws that they set up, they introduced and strengthened modern capitalist property relations. Helped by laws and economic relations conducive to it, market economy percolated into agrarian relations, dissolving the community relations it replaced. Ironically, the process picked up pace after Independence, land reforms creating a double edged process – they freed the tenants to participate more vigorously in markets, while they also forced the landlords who retained their lands to turn to wage labor and rich peasant farming to escape land reforms, in both cases extending and strengthening the reach of the market economy.

Water, however, has not followed the same trajectory. Surface water, theoretically and practically in cases of larger projects, has remained a state property, though groundwater has remained largely a private resource. Unlike countries like South Africa where state ownership is asserted as a trusteeship, a trust by the state held on the behalf of the community or the people, the post-colonial Indian state has asserted complete proprietorship over its water resources, mainly surface water, but, by implication, groundwater as well. With the coming of the wave of globalization and privatization, there is an increasing clamor for privatization in view of the apparent failure of state ownership as an instrument of water governance. However, as we have argued earlier, both concepts of state property and classical private property are not suited to tackling the special characteristics of water as an ecosystem and common pool resource, and we thus need to go back and learn from the community management forms that have worked so well in the past with respect to water.

Community bindings worked because they were based on consent and consensus. Traditional community bindings worked through traditional caste, tribal or clan obligations that created the consensus, though there is a need to recognize that they also carried within them the oppressive relationships and their associated paraphernalia. The consent and consensus was grounded in those oppressive relationships and were thus overthrown by those who sought to eliminate that oppression. Paraphrasing a warning by Prof. D. D. Kosambi, we would say that we cannot return to a golden age because there never was one; community and unity was always tainted by oppression and division. However, that does not mean that the past holds no lessons.

Given the high costs of exclusion – and these costs refer to non-consensual enforcement assessed in economic terms – pure economic measures are unlikely to work. We need to find new ways of mediation between the state and the individual that can form and support consensual forms of agreement. What is definitely needed is a process that brings together all the stakeholders/right holders in a process that will bring us as close to a consensus as possible. This needs to be a participatory process aided by administrative measures from the top. But for that to happen, we will have to look at water not as private property in the classical sense, but, as outlined earlier, as a share of common pool resource subject to priorities and norms with respect to use, and as rights and entitlements that carry with them responsibilities and obligations.

This means looking for new forms that will socially express and mediate these needs. For this, it is not sufficient to merely set up river basin authorities with centralized powers because they are bound to remain ineffective and their mandates unenforceable, or to merely take water out of the State List and place it in the Concurrent List as being suggested by some. We need to find new forms of participatory governance and new ways of evolving consensus; the old way of relying on the state to legislate, act and enforce on behalf of "the people" is much less likely to work if it is not supported by these processes. There is a need to look afresh at multi-stakeholder platforms or processes and participatory instruments for working out reasonable entitlements and arrangements for sharing during normal, surplus and shortage situations.

There is also a need to work out appropriate economic instruments that will adequately take into account the concept of access to water as a basic human right, as well as its aspect as private property. We need to see how our common pool of water resources can be utilized for ensuring the provision of basic rights with a high dependability without necessarily foreclosing options of economic use for the extra water that will be available in most years if we do so. Instituting graded tariffs with an assured minimum water service that constitutes a basic right to be available at an affordable rate combined with variable surpluses that may be provided at economic prices for remunerative and profitable use – is an attempt at bringing about this reconciliation. The important point to note is the need to be able to treat water in all its complexity – not only to be nurtured and prized if not worshipped as the basis of life, a basic right, and a social good but also to be utilized as a resource – productively, profitably and sustainably.

≈

Notes

1. It is true that some of the literature available globally, especially on the Middle East, sees water conflicts more in terms of wars, peace and survival. For example, see some of the representative works by Allan 1996; Myles 1996; Joyce 1995; Murakami 1995; Bulloch and Adel 1993. Periodicals like *Studies in Conflict and Terrorism* have published articles on water conflicts more or less from the same standpoint. Gleick 1993 discusses the history of water-related disputes and water resources systems as offensive and defensive weapons.
2. These conflicts are not very sharp in the developed countries of Europe or the US because these countries have evolved certain "social" consensus around a minimum set of benchmarks and norms. A certain institutional and legal space has also been provided for the settlement of water conflicts in these countries through negotiations and increasing presence of multi-stakeholder platforms and processes.
3. For a detailed discussion of this case, see Chauhan 2007.
4. On the other hand, it is also possible to treat – and it is so treated by many – the issue of allocation of water between *different* uses as an issue of equity, of a "fair" allocation between them.
5. For a detailed discussion on this, see Kavade-Datye 2007.
6. Dalits are also called the Scheduled Castes and were considered "untouchables" under the caste system.
7. For a detailed discussion, see Paranjape et al. 2007.
8. For details, see Jayakumar and Rajagopal 2007.
9. For details, see Janakarajan 2007.
10. For a detailed discussion on how to tackle the issue of water quality/pollution-induced conflicts, see Appasamy 2007.
11. For a detailed discussion on the conflict over Sardar Sarovar Project, see Sangvai 2007 and Paranjape and Joy 2007b.
12. This is important in the context of the polarized debate on "large versus small" or the "no dam" stand of some of the environmental movements. "In some drought prone areas, for example, in many areas in Maharashtra, millions of people are also uprooted due to drought and

the lack of supplementary irrigation from dams. They migrate to large cities like Mumbai and live in slums in the most degraded conditions. Secondly an extreme assumption that people would never want to move and should always stay where they have their roots also contradicts historical experience" (Patankar and Phadke 2007: 308).

13. For a detailed discussion on this possibility, see Paranjape and Joy 1995.
14. For a detailed discussion of the case, see Singh 2007.
15. For a detailed discussion on this idea, see Gleick 2002.
16. For details on Plachimada case, see Surendranath 2007.
17. For details on Sheonath river conflict, see Das and Pangare 2007.
18. For details of this approach, see Paranjape and Joy 1995; Datye et al. 1997.
19. For details of this system, see Datye and Patil 1987; GoM 1999; Sane and Joglekar 2007.
20. The issues related to integration of sources have been detailed in Paranjape and Joy 1995.
21. For details of the Ozar experience in integration, see Paranjape and Joy 2003.
22. "The existing water law framework in India is characterized by the co-existence of a number of different principles, rules and acts adopted over many decades. These include common law principles and irrigation acts from the colonial period as well as more recent regulation of water quality and the judicial recognition of a human right in water" (Cullet 2006).
23. "The overall supremacy and control by the state is sometimes loosely described as its "eminent domain," as Iyer (2009) puts it.
24. The fact that the primary entry relating to water is Entry 17 in the State List of the Constitution of India gives rise to the general impression that water is a state subject. That impression is only partly true. Entry 17 in the State List is not absolute: it is subject to the provisions of Entry 56 in the Union List which gives the center a role in relation to inter-state rivers to the extent that the Parliament can legislate on this subject. The fact, however, is that the Parliament has not used Entry 56 for the purpose of legislation to any significant extent. (There is indeed a River Boards Act 1956 under that entry, but as we will see, that Act has remained virtually inoperative.) (Iyer 2009).
25. "The lack of a comprehensive water legislation has ensured that, to date, water law is made up of different instruments, principles and judicial decisions which are not necessarily fully compatible with each other. This is further compounded by the fact that formal water law is supplemented by a number of customary and religious rules concerning water use an control whose application continues to-date in many places" (Cullet 2006: 206).

26. See Ramanathan (1992) for an account of the limitations on statute law placed by the fact that it is essentially about property and ownership and how this framework obtrudes even into those attempts of the state that are aimed at addressing inequity through a process of extension of bureaucratization and expropriation of power from the people and its concentration in its own hands, and consequently in the hands of the bureaucracy. This is compounded by the fact that the efforts of the state are indissolubly linked to the objective of revenue generation, which has meant that water rights and access have been mediated either through landed or industrial property.

References

Allan J. A. (ed.). 1996. *Water, Peace and the Middle East: Negotiating Resources in the Jordan Basin.* New York: St. Martin's.

Appasamy, Paul. 2007. "Water Quality Conflicts: A Review," in K. J. Joy, Biksham Gujja, Suhas Paranjape, Vinod Goud and Shruti Vispute (eds), *Water Conflicts in India: A Million Revolts in the Making*, pp. 137–44. London, New York and New Delhi: Routledge.

Bulloch, John and Darwish Adel. 1993. *Water Wars: Coming Conflicts in the Middle East.* London: Gollancz.

Chauhan, Malavika. 2007. "Biodiversity versus Irrigation: The case of Keoladeo National Park," in K. J. Joy, Biksham Gujja, Suhas Paranjape, Vinod Goud and Shruti Vispute (eds), *Water Conflicts in India: A Million Revolts in the Making*, pp. 14–20. London, New York and New Delhi: Routledge.

Cullet, Philippe. 2006. "Water Law Reforms: Analysis of Recent Development", *Journal of India Law Institute*, 48(2): 206–31. Also available online at http://www.ielrc.org/content/a0603.pdf (accessed September 3, 2013).

Das, Binayak and Ganesh Pangare. 2007. "A River Becomes Private Property: The Role of the Chhattisgarh Government," in K. J. Joy, Biksham Gujja, Suhas Paranjape, Vinod Goud and Shruti Vispute (eds), *Water Conflicts in India: A Million Revolts in the Making*, pp. 431–34. London, New York and New Delhi: Routledge.

Datye, K. R. and R. K. Patil. 1987. *Farmer Managed Irrigation Systems: Indian Experiences.* Mumbai: Centre for Applied Systems Analysis in Development.

Datye, K. R., with Suhas Paranjape and K. J. Joy. 1997. *Banking on Biomass: A New Strategy for Sustainable Prosperity Based on Renewable Energy and Dispersed Industrialisation.* Ahmedabad: Centre for Environment Education.

Gleick, Peter. 2002. *The New Economy of Water: The Risks and Benefits of Globalisation and Privatisation of Fresh Water.* Oakland, CA: Pacific Institute for Studies in Development, Environment and Security.

Gleick, Peter. 1993. "Water and Conflict: Fresh Water Resources and International Security," *International Security*, 18: 79–112.

Government of Maharashtra (GoM). 2005a. *Maharashtra Management of Irrigaiton Systems by Farmers (MMISF) Act*. Mumbai: Department of Water Resources, GoM.

———. 2005b. *Maharashtra Water Regulatory Authority (MWRRA) Act*. Mumbai: Department of Water Resources, GoM.

———. 1999. *Maharashtra Water and Irrigation Commission Report* (also called *Chitale Commission Report*), 5 vols. Aurangabad: Maharashtra Water and Irrigation Commission, GoM.

Habib, Zaigham. 2000. "Resource Conservation and Civil Society: Water and Food Security in Pakistan," in Peter Mollinga (ed.), *Water for Food and Rural Development: Approaches and Initiatives in South Asia*, pp. 184–96. New Delhi: Sage.

Iyer, Ramaswamy R. 2009. "A Synoptic Survey and Thoughts on Change," in Ramaswamy R. Iyer (ed.), *Water and the Laws in India*, pp. 567–623. New Delhi: Sage.

Janakarajan, S. 2007. "Conflict over Water Pollution in the Palar Basin: The Need for New Institutions," in K. J. Joy, Biksham Gujja, Suhas Paranjape, Vinod Goud and Shruti Vispute (eds), *Water Conflicts in India: A Million Revolts in the Making*, pp. 163–70. London, New York and New Delhi: Routledge.

Jayakumar, N. and A. Rajagopal. 2007. "Noyyal River Basin: Water, Water Everywhere, Not a Drop to Drink," K. J. Joy, Biksham Gujja, Suhas Paranjape, Vinod Goud and Shruti Vispute (eds), *Water Conflicts in India: A Million Revolts in the Making*, pp. 158–62. London, New York and New Delhi: Routledge.

Joy, K. J. and Suhas Paranjape. 2009. "Water Use: Legal and Institutional Framework," in Ramaswamy R. Iyer (ed.), *Water and the Laws in India*, pp. 213–50. New Delhi: Sage.

———. 2007. "Understanding Water Conflicts in South Asia," *Contemporary Perspectives*, 1(2): 29–57.

Joy, K. J., Biksham Gujja, Suhas Paranjape, Vinod Goud and Shruti Vispute. 2007. "A Million Revolts in the Making: Understanding Water Conflicts in India," in K. J. Joy, Biksham Gujja, Suhas Paranjape, Vinod Goud and Shruti Vispute (eds), *Water Conflicts in India: A Million Revolts in the Making*, pp. xvi–xxxii. London, New York and New Delhi: Routledge.

Joyce, Shira. 1995. *Covenant over Middle Eastern Waters: Key to World Survival*. New York: H. Holt.

Kavade-Datye, Namrata. 2007. "Tembu Lift Irrigation in the Krishna River Basin: Conflict over Equitable Distribution of Water," in K. J. Joy, Biksham Gujja, Suhas Paranjape, Vinod Goud and Shruti Vispute (eds), *Water Conflicts in India: A Million Revolts in the Making*, pp. 98–104. London, New York and New Delhi: Routledge.

Menon, Ajit, Pravin Singh, Sharachchandra Lele, Suhas Paranjape and K. J. Joy. 2007. *Community-based Natural Resource Management: Issues and Cases from South Asia*. New Delhi: Sage.

Mollinga, Peter, K. J. Joy and Suhas Paranjape. 2002. "Enhancing Productivity of Water for Poverty Alleviation and Livelihoods." Theme 2 paper presented at the Electronic Conference on Water and Food Security, organized by FAO, September 15–November 1, www.fao.org./landand water/aglw/wsfs/docs/theme2.pdf (accessed January 2002).

Murakami, Masahiro. 1995. *Managing Water for Peace in the Middle East: Alternative Strategies*. New York: United Nations University Press.

Myles, James R. 1996. *U. S. Global Leadership: the U.S. Role in Resolving Middle East Water Issues*. Carlisle Barracks, PA: US Army War College.

Narain, Sunita. 2007. "Privatisation: A Review," in K. J. Joy, Biksham Gujja, Suhas Paranjape, Vinod Goud and Shruti Vispute (eds), *Water Conflicts in India: A Million Revolts in the Making*, pp. 411–16. London, New York and New Delhi: Routledge.

Paranjape, Suhas and K. J. Joy. 2008. "An Alternative Approach to Drought and Drought Proofing," in Jasveen Jairath and Vishwa Ballabh (eds), *Droughts and Integrated Water Resource Management in South Asia: Issues, Alternatives and Futures*, pp. 98–122. New Delhi: Sage.

———. 2007a. "Equity, Access and Allocations: A Review," in K. J. Joy, Biksham Gujja, Suhas Paranjape, Vinod Goud and Shruti Vispute (eds), *Water Conflicts in India: A Million Revolts in the Making*, pp. 67–72. London, New York and New Delhi: Routledge.

———. 2007b. "Alternative Restructuring of the Sardar Sarovar Project: Breaking the Deadlock," in K. J. Joy, Biksham Gujja, Suhas Paranjape, Vinod Goud and Shruti Vispute (eds), *Water Conflicts in India: A Million Revolts in the Making*, pp. 338–43. London, New York and New Delhi: Routledge.

———. 2004. *Water: Sustainable and Efficient Use*. Ahmedabad: Centre for Environment Education.

———. 2003. *The Ozar Water User Societies: Impact of Society and Co-management of Surface and Groundwater*. Pune: Society for Promoting Participative Ecosystem Management (SOPPECOM).

———. 1995. *Sustainable Technology: making the Sardar Sarovar Project Viable*. Ahmedabad: Centre for Environment Education.

Paranjape, Suhas, K. J. Joy, Terry Machado, Ajaykumar Varma and S. Swaminathan. 1998. *Watershed Based Development: A Source Book*. New Delhi: Bharat Gyan Vigyan Samiti.

Paranjape, Suhas, Raju Adagale and Ravi Pomane. 2007. "Mahad to Mangaon: Eighty Years of Caste Discrimination: What Caste is Water?," in K. J. Joy, Biksham Gujja, Suhas Paranjape, Vinod Goud and Shruti Vispute (eds), *Water Conflicts in India: A Million Revolts in the Making*, pp. 111–15. London, New York and New Delhi: Routledge.

Patankar, Bharat and Anant Phadke. 2007. "Dams and Displacement: A Review," in K. J. Joy, Biksham Gujja, Suhas Paranjape, Vinod Goud and Shruti Vispute (eds), *Water Conflicts in India: A Million Revolts in the Making*, pp. 309–13. London, New York and New Delhi: Routledge.

Ramanathan, Usha. 1992. "Legislating for Water: The Indian Context." Working Paper, International Environmental Law Research Centre (IELRC), Geneva, Switzerland, http://www.ielrc.org/content/w9201.pdf (accessed September 19, 2013).

Sane, S. B. and G. D. Joglekar. 2007. 'The Collapse of Phad System in the Tapi Basin: A River Strains to Meet Farmers' Needs," in K. J. Joy, Biksham Gujja, Suhas Paranjape, Vinod Goud and Shruti Vispute (eds), *Water Conflicts in India: A Million Revolts in the Making*, pp. 89–96. London, New York and New Delhi: Routledge.

Sangvai, Sanjay. 2007. "People's Struggle in the Narmada Valley: Quest for Just and Sustainable Development," in K. J. Joy, Biksham Gujja, Suhas Paranjape, Vinod Goud and Shruti Vispute (eds), *Water Conflicts in India: A Million Revolts in the Making*, pp. 333–39. London, New York and New Delhi: Routledge.

Seckler, D., U. Amarasinghe, D. Molden, R. de Silva, and R. Barker. 1998. *World Water Demand and Supply, 1990 to 2025: Scenarios and Issues*. Research Report 19. Colombo. International Water Management Institute (IWMI).

Sen, Sumita. 2007. "The Indo-Bangladesh Water Conflict: Sharing the Ganga," in K. J. Joy, Biksham Gujja, Suhas Paranjape, Vinod Goud and Shruti Vispute (eds), *Water Conflicts in India: A Million Revolts in the Making*, pp. 403–10. London, New York and New Delhi: Routledge.

Singh, Vikas. 2007. "Struggle over Reservoir Rights in Madhya Pradesh: The Tawa Fishing Cooperative and the State," in K. J. Joy, Biksham Gujja, Suhas Paranjape, Vinod Goud and Shruti Vispute (eds), *Water Conflicts in India: A Million Revolts in the Making*, pp. 350–54. London, New York and New Delhi: Routledge.

Sinha, Rajesh. 2007. "Two Neighbours and a Treaty: Baglihar Project in Hot Waters," K. J. Joy, Biksham Gujja, Suhas Paranjape, Vinod Goud and Shruti Vispute (eds), *Water Conflicts in India: A Million Revolts in the Making*, pp. 396–402. London, New York and New Delhi: Routledge.

Surendranath, C. 2007. "Coke Versus the People at Plachimada: The Struggle over Water Continues," in K. J. Joy, Biksham Gujja, Suhas Paranjape, Vinod Goud and Shruti Vispute (eds), *Water Conflicts in India: A Million Revolts in the Making*, pp. 419–24. London, New York and New Delhi: Routledge.

World Commission on Dams. 2000. *Dams and Development: A New Framework for Decision Making*. London and Sterling: Earthscan.

9

Inter-sector Allocation of Hirakud Dam Water

An Economic Analysis

Sanjukta Das

During the last few decades, water conflicts have increased all over the world. They are mainly focused on the inter-sectoral allocation of water and differential access to water within the sector. Water is used for household consumption (i.e., drinking, cooking, cleaning, bathing, etc.), as well as for production in agriculture and industry. Economic growth, rising industrialization and urbanization have increased the demand for water over time. Water, categorized as a free good in the Higher-Secondary-level[1] text-books on economics, has become scarce in relation to its multiple and multisectoral uses. However, unequal property rights of access to and use of this scarce resource has provoked a number of conflicts among its users.

In the past, multipurpose river valley projects were undertaken to control floods and use water for irrigation, electricity generation and other purposes. The distribution of water stored by big dams for different purposes has become very complex at present. Since the governments are responsible for the construction of these big dams (because of the huge costs and long gestation period involved), they also assume the responsibility for water management in many cases. Over time, priorities attached to different purposes for water use and, accordingly, the allocation of water have very often changed. These changes have created a lot of discontent among the affected parties which now demand a review of the water policies.

In the 1950s, the Hirakud Multipurpose River Valley Project was built on the river Mahanadi in Sambalpur, Orissa (now Odisha), India. Hirakud dam is the longest river dam in the world with a length of 25 km. For this, 74,057.47 hectares (ha) of land in the undivided Sambalpur district[2] was submerged by the Hirakud reservoir, of which 49,776.33 ha was fertile agricultural land. The dam was constructed mainly to control floods, irrigate large areas of western Orissa and generate electricity. Promotion of pisciculture and the provision of drinking water to the Sambalpur municipal areas and its agglomerates were some of the other objectives. Admittedly, the dam has reduced the severity of floods, become the main source of hydropower in the state and served the region by providing irrigation water and thereby helping the rise in agricultural output. However, it has also been supplying water to the nearby industries though in the original plan there was no provision for the use of dam water for industrial purposes.

Over time, with the increase in demand for water from different sectors, there has been a growing pressure on this resource. Apart from the change in state government's water policy in the favor of industrial use of water, tension and discontent among water users for irrigation water in the area has been reflected in their organized protests.[3] On October 26, 2006, 20,000 farmers of western Orissa staged their token opposition by creating a human chain from Gandhi Minar to Nehru Minar of the Hirakud dam. When this opposition failed to create any favorable policy change, they resorted to a civil disobedient movement on November 6, 2007. Some 30,000 farmers gathered for this purpose at the Hirakud dam.[4]

This chapter, written against the backdrop of this competition and conflict between multiple uses of water, discusses the intersectoral allocation of Hirakud reservoir water. The first section discusses the supply situation of the water from the reservoir. The second section analyzes the different uses of the reservoir water. The third section throws light on the Government of Orissa's recent water policy and its impact on Hirakud dam water management. The fourth section discusses the issues relating to inter-sector water allocation. The fifth section discusses the issues of efficiency and equity in the irrigation system. In the next section, ecology and environmental issues are discussed. In the final section, an attempt is made to examine the ways of conflict resolution.

Availability of Water in the Hirakud Dam

The maximum water level of the reservoir which is otherwise known as the full reservoir level (FRL) is 192.024 m (630 ft) out of which the dead storage level (DSL) is 179.830M (590 ft).When the water level is below the DSL, it cannot be used.

During the years of bad monsoon like 1966, 1977, 1979 and 2000, the water level in the reservoir was very low. However, the volume of water in the reservoir depends not only on rainfall, but also on the inflow of Mahanadi water to the reservoir from the upper catchment of Hirakud, and on the release of water from the reservoir during monsoons by the dam management authority to protect the dam without causing floods. Thus, the quantum of reservoir water becomes variable at different points of time. Figure 9.1 presents the water level of the reservoir in different years. As can be discerned from the figure, the maximum water level was very low in 1966, 1977, 1979 and 2000. In many years, especially from 1972 to 1988, the minimum water level was below the DSL.

The low water level in the reservoir may be attributed to meager rainfall, less inflow of Mahanadi water from the upper catchment of the dam and increasing siltation in the reservoir. According to an estimate, the dependable water flow to Mahanadi was 25 million acres foot (MAF)[5] in 1950 which declined to 16 MAF in 2000. This fall was mainly due to the large number of dams constructed on the tributaries of Mahanadi in Chhattisgarh. Moreover, the water holding capacity of the reservoir has been falling at an annual rate of 0.4 percent. In the beginning, the useable water (live storage capacity) of the dam was 4.22 MAF, which declined to 4 MAF in 1995 as per the Central Water Commission (CWC) Report (Mohanty 2008). Rathore et al. (2006) found that the live storage loss in sedimentation was 984 M m^3 (16.9 percent of live storage). According to the Technical Committee of the State, also known as the Jeyaseelan Committee (Government of Orissa 2007a), it had fallen to 3.77 MAF by 2007. Soil erosion by the tributaries of

Map 9.1
Location of Hirakud Dam

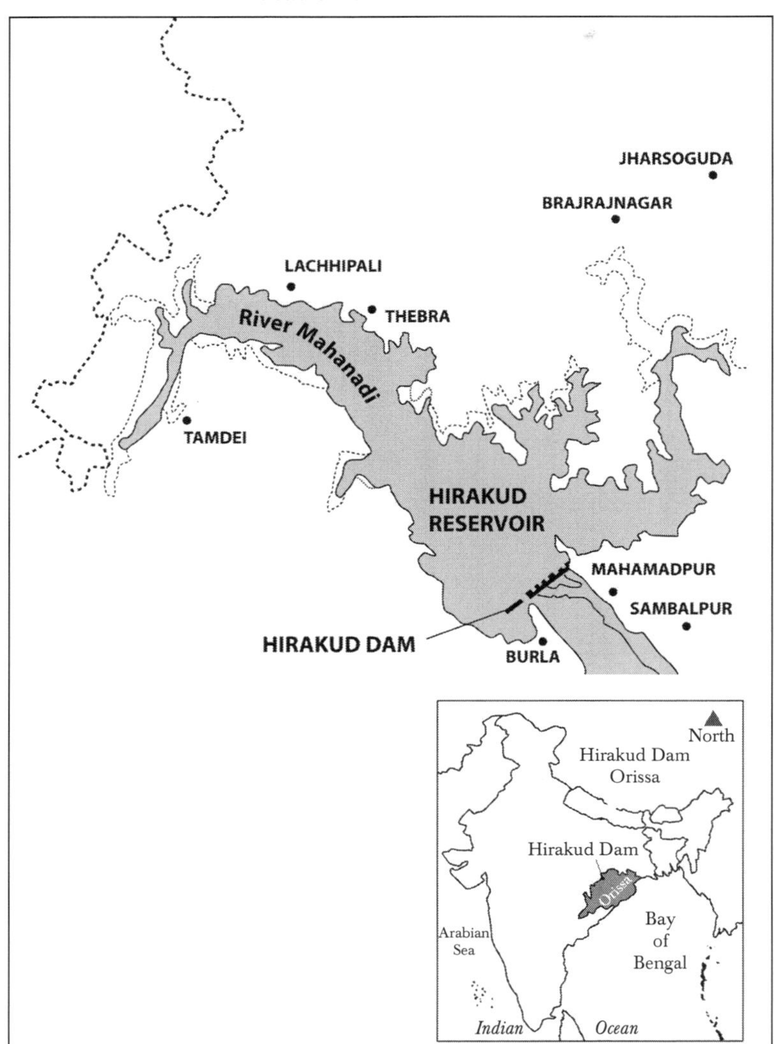

Source: Prepared by the author. Map not to scale.

Mahanadi and creation of Mahanadi delta are some of the causes of siltation. All these factors have reduced the supply of water to the dam for its subsequent use.[6]

Figure 9.1
Year-wise maximum and minimum reservoir level with respect to FRL and DSL

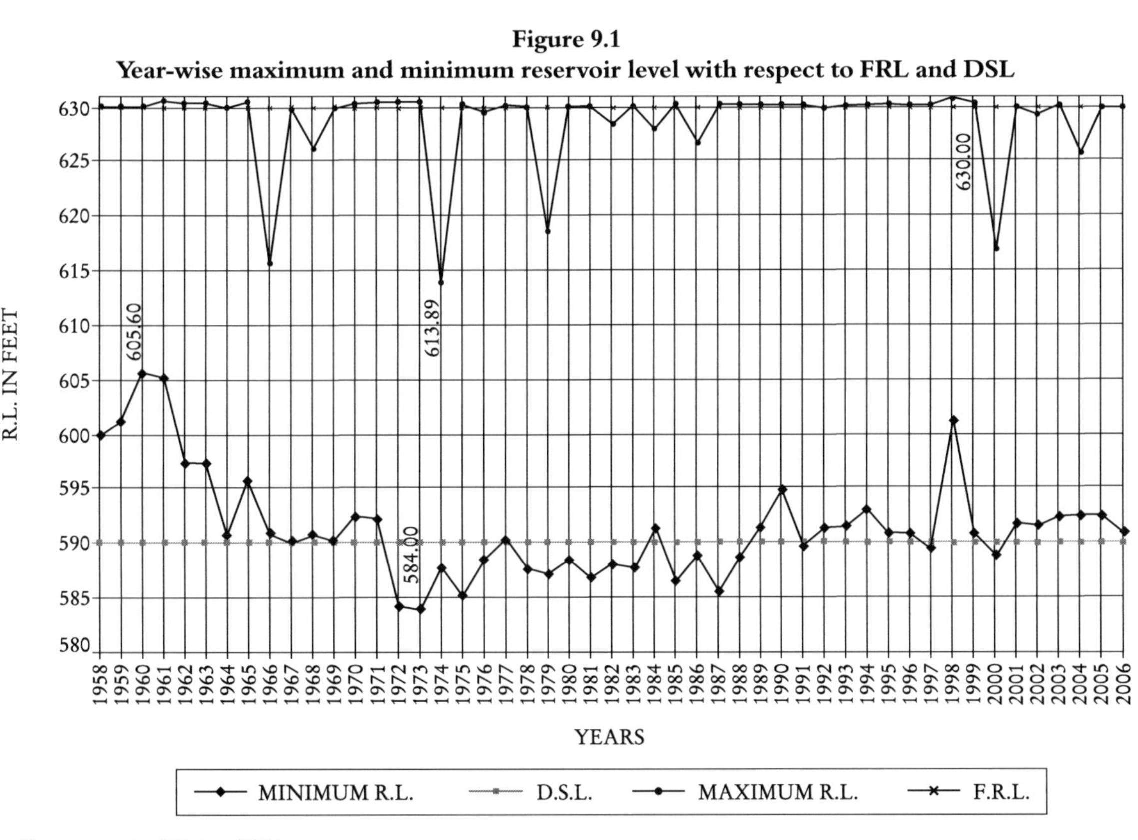

Source: Government of Orissa 2004.

Uses of the Reservoir Water

The different uses of the multipurpose Hirakud dam are: hydropower generation, irrigation, supply of water to houses for drinking and other household purposes, pisciculture, maintenance of ecology and supply of water to nearby industries. The state government, as manager of this resource, allocates water for these purposes following its own guidelines. As supply of water has been falling over time, efficient and equitable allocation of the available water among the competing demands has been a big challenge for the government.

Hydropower Generation

Hydropower is generated from the water in the reservoir through two power houses, one at Burla and another at Chipilima, with the installed capacity of 307.5 MW. A 25 km power channel carries tail-race[7] water from Burla to Chipilima power house. The former has seven power-generating units and the latter has three. The power generation in the first power house started in 1956 and in the second from 1962.

Figure 9.2 presents the statistics for year-wise (July–June) hydropower generation. It can be seen that over time, electricity generation has increased (reaching the peak, i.e., 1307.16 Million Units [MU] in 1990). However, power generation was very low in some years like 1966, 1974, 1979, 2000 and 2002.[8] Incidentally, these are also years of deficit rainfall (Government of Orissa 2004). Hydropower generation, of course, depends on the water released from the reservoir for this purpose and the number of power units in operation. The former, i.e., water released from the reservoir, depends on the water stored in the dam and the quantity of water the authority decides to release.

Irrigation

There are three canals, i.e., Bargarh canal, Sambalpur distributary and Sasan canal which take water from the reservoir. These canals have branch canals, distributaries, minors and sub-minors.[9] The

Figure 9.2
Power generation (in MU) from Hirakud

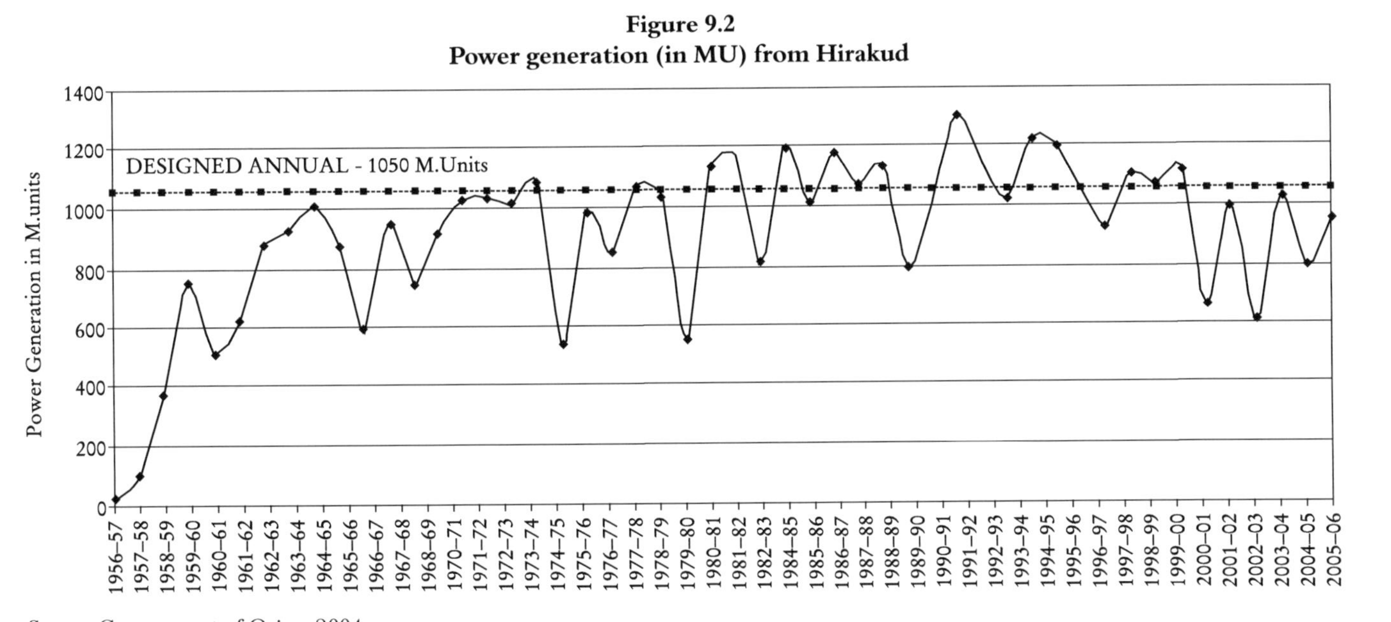

Source: Government of Orissa 2004.

irrigation practice in the command area is demand based.The water supply in the channel is on a continuous basis, and the farmers irrigate their lands as per their requirement by closing and allowing the water (Nayak 2006). In the entire Hirakud command area, paddy is the predominant crop covering 95 percent of the total crop area (NRSA 2004). The canals irrigate 159,100 ha in *kharif* and 106,820 ha in *rabi* seasons (Government of Orissa 2007). These figures are higher than the estimates in the original first-phase plan (153,750 ha for *kharif* crop and 76,875 ha for *rabi* crop) of the dam, i.e., of 1953. As per the information based on satellite images in 2006–07, 163,036 ha of *kharif* crop and 115,481 ha of *rabi* crop land were irrigated by the canals of Hirakud dam (NRSA 2004). It is quite natural that in course of time the area under irrigation has increased and many unoccupied fallow government lands, communal lands and even forest areas in the vicinity have been converted into crop lands because irrigation facilities have been made available.[10] In addition to water from canals, water released in course of hydropower generation is utilized to irrigate 167,000 ha of *kharif* and 100,960 ha of *rabi* crop in the Delta Stage-I, and 136,000 ha and 67,622 ha of *kharif* and *rabi* crop respectively in the Delta Stage-II area (Government of Orissa 2007). All these have increased the area under production and thereby the output in the area.

Over time, the demand for water for irrigation has increased. This demand depends on the crop calendar, cropping patterns and the cropping intensity. Change in the cropping pattern and the introduction of high yielding variety (HYV) of crops change the irrigation demand. Meher and Baboo (2003) have shown that there is water scarcity in the Hirakud command area which has reduced the area under cultivation especially in the *rabi* season. This finding is contrary to the official statements that point to a rise in the irrigated area. In the command area, there is also increased mono cropping (to paddy) and introduction of HYV seeds (D'Souza et al. 1998; Baboo 2009). Using the remote sensing method, Nayak (2006) has estimated the water demand in parts of Hirakud command area and found shortage of canal water. Raul et al. (2008) have also found that the present rice-based cropping pattern of Hirakud command area has resulted in the demand for water exceeding the supply, especially in the *rabi* season.

Government of Orissa (2007) has admitted that the dependable inflows during non-monsoon months are critical to meet the targeted water requirements of various sectors including irrigation. The amount of water available in the field, however, depends on the amount of water released from the dam (which, again, is a function of water availability in the reservoir). It also depends on factors like loss of water on the way due to evaporation, leakages, methods of taking water from canals to fields (i.e., flooding or water courses), location of the field in relation to the main canal (i.e., at the head-end or tail-end), distance of the field from the minor canal, cropping pattern adopted by the upper-end farmers, etc. For instance, the less efficient canal flow method of water transfer to crop lands, poor maintenance of canals (Meher and Baboo 2003) and evaporation[11] of water on the way have reduced the supply of water to the fields. As a result of all these (most probably), canal water is not reaching the fields located at the tail-end. The report (2003) prepared by the Development Support Centre (DSC) for the Planning Commission of India has expressed the concern that while all the irrigated land of the head-end (upper) of the Hirakud canal are getting water; only 35 percent of the middle-end irrigated land and 18 percent of the tail-end irrigated land of the canal are receiving water.

Pisciculture

Fishing activities in Hirakud are under the purview of cooperative pisciculture societies. The Fisheries and Animal Resources Development Department of the state government drops fish seeds, imposes restrictions on the release of waste water by industries into the reservoir, enforces stringent rules on the use of wide fishing nets for fishing and restricts fishing during the breeding season. Mohanty (2008) has estimated that approximately 2,500 people earn their livelihood from pisciculture in the Hirakud dam. However, over time, the fish production has fallen (see Table 9.1). The per-head catch has declined from 524 kg in 1980–81 to 239 kg in 1990–91 (Fisheries and Aquaculture Department n.d.). Moreover, during the last 50 years, the number of species of fish in the dam has declined from 104 to 43. Increase in the population of

Table 9.1
Fish output from Hirakud Dam

Year	*Output (in Metric Tonne)*
1980–81	843
1990–91	332.8
1995–96	358.6
1996–97	267.3
1999–2000	259.24
2003–04	163.78
2004–05	151.54
2005–06	182
2006–07	205
2007–08	217.53
2008–09	245.15

Sources: Mohanty 2008; http://newindianexpress.com/states/odisha/article142635.ece (accessed September 16, 2013).

fishermen and non-availability of alternative and better livelihood opportunities to them, operation of fish mafias, illegal fishing in violation of the rules set by the Fisheries Department and the use of closely knitted fishing nets by the traders – are the probable causes for the decline in the output (i.e., catch).

Over and above these causes, the disposal of industrial waste water is also said to be responsible, to a large extent, for loss of fish species and volume. The fishermen complain that the waste water of different industries when released into the reservoir or feeding rivers results in a typical disease, *kshata rog*, among the fishes (Pati and Biswal 2009).

Drinking Water

Safe drinking water is one of the basic needs of the people. Mahanadi water is supplied as drinking water to the Sambalpur municipal areas and its agglomerates. This river also acts as the urban liquid waste sink of this town. The Sambalpur municipality draws water from the Hirakud reservoir[12] and supplies drinking water to the town. However, many areas of Sambalpur town do not get adequate water, even if they are connected with pipes carrying drinking water. Moreover, the construction of the dam has reduced the water flow into the river when it passes through the Sambalpur town. Increased waste inflow and reduced water flow (especially from November to June) makes the water quality very poor and unsuitable for drinking and organized outdoor bathing. In Sambalpur, near downstream areas, Mahanadi water is suitable only for pisciculture, wildlife habitat, irrigation, industrial cooling and controlled waste disposal.[13] A suggestion to release 100 cusecs[14] of water once in a week during these dry months (April, May and June) was made by the then Chief Engineer M. S. Thirumale Iyenger at the beginning of the dam operation. However, this suggestion has not been implemented till date (Mohanty 2008).

Industrial Use of Dam Water

At the project report stage there was no explicit provision for water allocation for the industry.[15] However, from the very beginning of the dam operation, industries were set up, one after another, in the neighborhood of Hirakud. INDAL (Indian Aluminum Plant) started drawing water from the Hirakud reservoir from 1959; gradually, Tata Refractories of Belpahar and Orient Paper Mill of Brajarajnagar (now closed) too followed the suit.[16] During the 1960s and 1970s, there was no conflict over the use of dam water. However, over time, the number of plants in the neighborhood of Hirakud dam, as well as that of the rivers Ib and Veden[17] has increased. According to the Government of Orissa report (2007a), 15 industries are permitted to draw 0.36 MAF water per annum from the Hirakud reservoir and five more have applied for the release of 0.316 MAF water per annum from the reservoir. Mahanadi river basin met 50 percent of the water demand of Orissa in 2001 by providing 267.5 million m^3 of water (Government of Orissa 2004).

Government Policies

The state government's policies regarding water allocation and management of Hirakud are part of its larger policies of natural resource management. They have also been influenced by the broader policy frameworks that the national and the state governments have adopted from time to time.

In the 1950s, when the country initiated planned economic development, basic and key industries were established for industrial development. Big multipurpose river valley projects were undertaken for increasing agricultural production and electricity generation for domestic and commercial uses. These were inward-looking policies that were aimed at achieving self-sufficiency for the country. At the time of the planning and construction of Hirakud dam, the communities in the area were mentally prepared for large-scale displacement. The rural communities were given the assurance that the dam would bring prosperity to the rural areas in the vicinity through a positive transformation of the agricultural landscape and rural electrification. Maintenance of the dam and its canals was the responsibility of the government. The management of its water was also kept under the government's purview in order to promote equity of access. Water was allocated mainly for irrigation and hydroelectricity generation. It was made available to the farmers on the basis of the riparian principle,[18] and they paid water charges at subsidised rates. Since this dam whose construction involved large-scale displacement of communities and loss of arable land was constructed for creating irrigation facilities and generating electricity, allocation of the dam water for industrialization was not at all a priority at that time.

In 1990s, India adopted the Structural Adjustment Program (SAP), which sought to[19] achieve higher economic growth by promoting productivity, competition, greater market access and efficient utilization of resources. New industrial policies were announced; more and more concessions and facilities were offered by the Government of India, encouraging the private sector to set up industries. Since the neighborhood of Hirakud has rich mineral deposits of iron-ore, coal and bauxite, the liberalization policy of the government encouraged more and more mineral extraction and the establishment of mineral-based industries. This also raised their requirements of water. In 1990, the Government of Orissa earmarked 0.350 MAF (9 percent of the total usable water from the dam) from the reservoir for use by industries (vide).[20] This move was meant to encourage industrialization in the state. The state government also stated in the letter that there was surplus water in the reservoir and hence 10 percent of the water could be given to the industries. However, the government estimate of the surplus water is not substantiated.

The year 1997 marked an important development in the history of water resource allocation among different sectors, when the state government renamed the Irrigation Department as Department of Water Resources (DoWR). This indicated a marked shift in the attention paid by the government to this resource. Water resources meant for irrigation was to be shared by different competing users like industry to raise the productivity of water. However, the value of increase in industrial output is to be compared with that in agricultural output in order to com-pare the productivity of water use in agriculture and industry. Orissa Water Planning Organization (OWPO) was entrusted with the preparation of macro-level, multisectoral River Basin Plans (RBP) keeping in view water allocation priorities set out in the policy document of 1994. River Basin Organizations (RBOs) were established to plan and manage the water resources of different basins. Plans prepared by the OWPO had to be accepted by the RBOs; thereafter they could be approved by the state-level Water Resource Board. However, water allocation to industries increased manifold within a short span of time. It has been estimated that water allocation from Hirakud reservoir to the industry was 3,191.2 million gallons[21] per year (LGY) till April 1997 and it increased by 27 times in a period of nine years (Panda 2007).

Government of India's new water policy of 2002[22] was in line with the industrial policy of the 1990s. Orissa also prepared a similar draft water policy in 2003, wherein priority was given to the allocation of water for drinking first and then for irrigation. But one section of the draft policy stated that the priorities could be modified or added if warranted by area-specific considerations, though no change would be made in the priority given to allocation of water for drinking. This created an apprehension among the farmers that water allocation for irrigation might lose its priority in future. Similarly, as per paragraph 16.1 of the draft, private sector was to be allowed to own, operate, lease and transfer water resources, though without taking away local people's riparian and traditional rights over water. The earlier water policy of 1994 did not have any such provisions for water management by private sector. The new draft policy thus faced criticism from people, especially farmers, and even organized protests. However, the State

Water Policy of 2007 adopts the following areas of inter-sectoral water allocation in the order of priority: (*a*) water for drinking and other domestic uses (human and animal consumption), (*b*) ecology, (*c*) irrigation, agriculture and other related activities including pisciculture, and (*d*) hydropower, industries including agro-based industries and navigation and other uses such as tourism. It also categorically states that any alteration in the priorities would demand formulation of a new policy.

The state government considers this policy as a means to ensure financial sustainability and promote efficiency in water use, employment and economic growth. However, the farmer organizations (FOs) see it as an attempt to marketize dam water and facilitate the capitalists' access to and control over water, which would, in turn, lead to unemployment and livelihood insecurity for the poor in the locality. The FOs further argue that the usage rights of local agrarian communities over the dam water are being violated by the government's provisioning of water to the industries, more so when many command areas do not get water. Faced with stiff and massive resistance from farmers (especially after November 26, 2007), the Chief Minister of Orissa declared that not a single drop of water would be given to the industry.[23] However, no policy measure has been taken so far in this regard. In fact, by 2007, 22 industries had got the permission to use Hirakud reservoir water amounting to 478 cusecs (Mohanty 2008). Another industry, viz., Reliance Energy is awaiting permission. It is alleged that the industries would draw more water than stipulated in the memoranda of understanding (MoUs) with the government using various tactics because their requirements exceed their entitlements.[24] It is also alleged that without carefully examining the water requirements of the industries, the government has signed the MoUs. There are instances of gross violation of government order, of course, implicitly by the industries.

Issues Relating to Inter-sector Water Allocation

Growing competition between agriculture and industry for the use of water from the Hirakud dam and the grass-roots movements spearheaded by FOs against the government policies has raised issues related to the efficient use and equitable distribution of a scarce resource like water, as also the environmental consequences of the different uses of the dam water. It has been argued that efficiency in water use can be achieved when per unit yield is equalized across sectors. If this requires the transfer of water from agriculture to industry, it may promote efficiency but at the cost of inequity and food insecurity. This inequity can be redressed if the losers are compensated by the gainers and efficiency in the agricultural use of water is improved.

The competition and conflict between agriculture and industry over the use of Hirakud dam water is not unique. Such cases are found both in India and abroad. For instance, in Palakkad district of Kerala the extraction of groundwater by a multinational soft drink company has depleted the aquifer, dried up several wells and caused serious environmental damage (Molle and Berkoff 2006). Similarly, in the outskirts of Mumbai, the same company extracting groundwater to tap the fast growing market for bottled water used by the middle class of the city has provoked protests by farmers (Gandy 2005). Instances of such conflicts have also come to the fore in other countries like China, Thailand, Yemen and Pakistan (UNDP 2006). The challenge, therefore, is how to manage the transfer of a share of water resources from agriculture to non-agricultural users without doing injustice to the former because the diversion can have profound impact on the livelihoods of many, even if the volume transferred to the industry is less than that used in agriculture. For instance, only about 6 percent of the total water in the Hirakud reservoir is transferred to the industry.[25] In contrast, approximately 27 percent of the reservoir water is available for canal irrigation and approximately 59 percent for electricity generation (Mohanty 2008). The water used for electricity generation is further used for pisciculture, irrigation (lift irrigation in western Orissa and canal irrigation in Mahanadi delta area) and for other purposes. However, water used by the industry is not available for any other purposes.[26] If the loss is calculated in terms of the loss of water for irrigation and livelihood security of farmers, it would amount to the loss of water for irrigating 47,800 acres of land. According to a conservative estimate, 30 percent water loss on the way amounts to a loss of 33,500 acres of land that could otherwise be available for irrigation. This area of land could provide livelihood security to approximately 11,000 households with an average family size of five. Similarly, the shortfall in electricity generation per annum would amount to 32.026 million units because of the diversion of 478 cusecs of water from hydropower generation to industrial use; this loss in monetary terms would amount to INR 32.026 million, if electricity consumption were to be charged at the rate of INR 1 per unit (Government of Orissa 2007a, paragraph 6.2.5). However, the government has argued that the gap in hydropower generation can be bridged by thermal power generation. But the cost per unit of thermal power is more than twice that of hydro-electricity (Mohanty, 2008). Moreover, thermal power generation also requires water and at the same time pollutes the environment. Further, the diversion of water from power generation to industry would reduce the area under irrigation in the Mahanadi delta and the water flow in the Mahanadi river during non-monsoon months. In Orissa, water scarcity induced by variation in rainfall is quite common. The unpredictable variability in the water level of Hirakud dam due to variation in rainfall becomes a threat to the livelihoods of a large number of farmers and agricultural laborers. Under these circumstances, the diversion of water from agriculture to industry can, therefore, trigger rural poverty and malnutrition through reduced employment, income and food production in western Orissa where magnitude and density of poverty is already high.

As competition for water has intensified, development and use of market-based tradable water rights to resolve the rising demand may be considered as a remedial strategy. It would ensure the flow of water for its most productive uses. Experiences in other countries (e.g., USA, Chile, etc.) reveal that it has enhanced efficiency in water use. But this efficiency in water use has taken place in many cases at the cost of deteriorating equity situation. For instance, water trading in the western US has resulted in an aggregate increase in public welfare, but has caused losses to a large group of poor farmers, among others (Villarejo 1997, in Meinzen-Dick and Ringler 2006). In another instance, the development of water market in Chile has enhanced efficiency in water use, but these private water markets have undergone a process of complex interactions fraught with tensions with regard to bringing about efficiency and equity. Agriculturists and water-intensive industries such as mining have responded to higher water prices by adopting new technologies, including the drip irrigation systems that have created and sustained a boom in the export of high-value fruits and vegetables. But in the beginning, the deployment of these technologies had led to water monopolies, market distortions and highly unequal outcomes. To regain their lost rights to water resources, indigenous groups had to mobilize themselves and fight against such water monopolies. Finally, the Water Code Reform was adopted by the Chilean government in 2005 to address the problem.

In Orissa, the most favorable situation for a farmer is to have user rights over the Hirakud canal water. For the implementation of tradable water rights, institutional arrangements are to be made. But the apprehension that remains is that even if institutional arrangements are made, they may not be successful in the face of a high level of illiteracy among farmers and the consequent asymmetry of information between farmers and industries. Moreover, in spite of all these problems, even if tradable water rights are introduced, the possibility of tension and marginalization of the poor cannot be avoided. The possibility of the sale of water usage rights by farmers under duress or distress and force might become a common phenomenon and cause further disempowerment and livelihood insecurity for farmers. However, tradable water rights are feasible in case of intra-village transfer of water rights among the farmers. But it would not be a means to aid inter-sectoral allocation of water. This is so because as long as the riparian principle is in place and irrigation gets priority over industrial water use, a farmer will not be able to transfer his rights to an industrialist. According to this principle, an industrialist can get water right from the farmers when there is no demand for water for irrigation. This is, however, an unrealistic assumption.

The transfer of water from agriculture to industry can also take place through administrative fiat, market exchange or other types of negotiation. The transfer of water and the nature of water management will depend upon the nature and importance of water rights and the relative power and influence of different stakeholders. In Orissa, water transfer from Hirakud dam to industry takes place through government order. Of course, it is influenced by the government's industrial policy and the industry's intense lobbying for access to water. However, licenses and permits designed to facilitate adjustment to the growing competition may be far from equitable as can be seen in China and Philippines.[27] In fact, what has happened in China and Philippines can also happen in Orissa with regard to the use of Hirakud dam water. In the low rainfall years when the need for irrigation water is the highest, the available scarce water (because of low rainfall) could be siphoned off by the industrial users by exercising their influence over the government. This would deprive the poor farmers of the much-needed irrigation water. Under unequal power relationship, together with formal licensing system, thus, would thwart the goal of achieving equity in water distribution and use.

As has been argued, formal water rights offer no guarantee of equity in the face of unequal power relations. But at the same time it is to be admitted that the absence of a well-defined, properly regulated and enforceable rights framework is even less likely to enhance water security and more likely to open the door to institutional "water grabs" based on power. Formal water rights if regulated well, can play an important role in bringing about positive outcomes from the inter-sectoral transfer of water. Strong rights enforcement mechanism can help vulnerable producers resist the encroachments by large industries, commercial agriculture and urban users. In Orissa, granting such rights can also help the farmers reduce the problems arising out of the diversion of water to non-agricultural uses. Thus, the more feasible option is to gradually develop the existing water rights and strengthen the provisions for their exercise by the poor.

Efficiency and Equity within the Irrigation System

There is inefficiency in water use both in the industrial and agricultural sectors. Both are getting water at subsidized prices. Easy availability of raw materials (mineral ore) in the locality, and supply of cheap (sometimes subsidized) energy and dam water by the government has encouraged the establishment and operation of industries in this area. In this situation, it cannot be claimed that there is efficiency in industrial water use. In this section, we, however, focus on the efficiency in agricultural water use and examine the possibility of water transfers to the industry without the loss of agricultural output and aggravation of food insecurity.

An important fact to be noted is that the tail-end farmers suffer from twin disadvantags: less water and more uncertainty of getting water at the time of need. In India and Pakistan, it is typical for the tail-end farmers to get less than one-third of the water available to farmers at the head end (Hussain 2005). This is also true of the water use pattern in the Hirakud dam command area. The head-end farmers, encouraged by relatively better water security, adopt the HYV cropping pattern, cultivating water-intensive crops like rice, vegetables, etc. This practice reduces the water available to the tail-end farmers who are, in turn, forced to reduce their cropping intensity. Compounding this inequity is the fact that all users (head end and tail end) are charged a fixed rate for water use, irrespective of the volume of water used by them. At present, there is no mechanism to achieve equitable water distribution among head- and tail-end users. Such equity can be conceived through cooperation among the users. However, achieving equitable distribution of water is also subject to the effectiveness of local water users' associations (WUAs) in framing their rules regarding the sharing of water among their members, as also in enforcing them and monitoring their operation. Equitable distribution of irrigation water among different end users and transfer of water from the head-end to the tail-end users can improve the overall agricultural scenario. Such equitable distribution of water can influence the choice of cropping patterns and a likely preference for less-water intensive ones. However, this significant point is not articulated by the social activists at the grass roots level who lead the movement in favor of the farmers. Even the tail-end farmers do not seem to highlight this promising remedy to the problem of inequitable access to water. The explanation for this silence, however, lies in politics and not in economics. The rich head-end farmers do not want this issue to be highlighted; rather, they argue that over time, decline in actual irrigated land at the tail end would be a consequence of government's provisioning of water to the industry. On the other hand, both the government and the activists point to the poor maintenance of canals and the falling water holding capacity of the reservoir as some of the reasons for less water availability to the tail-end farmers.

The present system of dam water use in agriculture through canals is not efficient. It incurs a huge loss of water on the way to the field. The flooding method of irrigation not only requires more water per sq. ha but also creates problems like water logging and salinity, especially at the head-end of the canal. Soil survey department of Sambalpur found, from a survey in 1974, that out of 51, 921 ha of land, 2,927 ha is water logged (D' Souza et al. 1998). Panda and Nayak (2005) have found that (in total) 13 million ha of land has become unfit for cultivation because 6 million ha is water logged and 7 million ha is saline. Drip irrigation can help save water for irrigation, increase water-use efficiency, decrease the need for tillage, improve crop quality, and increase crop yields and fertilizer-use efficiency (Sivanappan 2002; Namara et al. 2005). Further, the technique is free from the problems of water logging and salinity (Narayanamoorthy 1997). Experience shows that the use of drip irrigation can lead to reduction in water use by 30–70 percent and increase crop yields by 20–90 percent (Postel 1999). In the opinion of irrigation experts, with the use of drip irrigation, even the water requirement for channels constructed to carry water directly to the field from the minor canals can be reduced by 30 percent. Admittedly, initial expenditure for setting up drip irrigation infrastructure and water courses is very high and farmers may not be able to afford. However, if industries are willing to provide necessary finance to build the infrastructure in order to get water in lieu, agricultural output can be maintained without further demand for water, and water can be made available to the industries.[28] But all these would only be possible when strong WUAs are present with whom the industries can negotiate. WUAs can also increase the bargaining power of poor farmers and protect their interests.

Siltation and poor maintenance of canals is another reason for inadequate water supply for irrigation. Lack of government funding is said to be the cause of their poor maintenance. Present water charges do not reflect the scarcity value of water and this, in turn, leads to wastage. The charges do not cover even the operation and maintenance (O&M) costs (ADB and DoWR 2010). Fixing rates so as to meet, at least, the O&M costs from the collection of water charges will ensure the sustainability of the system. But the estimation of cost and fixing of water charges must be transparently done and, if possible, with the participation of water users so that they are informed of and less resistant to rate hike. This will also reduce the wastage of water. Similarly, differential charges based on the ability to pay and the service received would not only raise more funds but also be equitable. Cooperation and co-ordination among the user groups (achieved through dialog) can bring about equitable water distribution and efficient water use within the irrigation system. There is also a need for forming a sub-group of beneficiaries from within the head-end and tail-end users. Mobilization and group formation of the latter is more essential than that of the former. Imposition of economic prices for water use or withdrawal of irrigation subsidy is likely to invite a lot of criticism and resistance initially, but such resistance can be overcome if the amount collected would be returned to the rural areas honestly and transparently and, in turn, used for rural development and employment generation. Sometimes, the industries' initiative, although guided by their own cost–benefit analyses, to bear a part of the maintenance costs can also reduce the ill-feeling amongst the farmers towards the diversion of water to industries.

Ecology and Environmental Issues

Construction of Hirakud dam on the Mahanadi river for multipurpose, inter-sectoral uses of water is not free from environmental consequences. As mentioned earlier, control of water through the dam has resulted in poor water quality especially in Sambalpur town during the non-monsoon months, causing environmental and ecological problems. The negative impact on aquatic (both plant and animal) life is also a matter of concern which one cannot ignore.

The water quality of Mahanadi was studied for the period from 1983 to 1988[29] and the results of this study are frequently cited to express concerns regarding the quality of water in the river. Recently, waste disposal into Mahanadi from the surrounding industries is expected to have adverse effects on water and aquatic life. Decline in the fish varieties in the river could be one of its consequences. However, in the absence of any empirical studies, it is difficult to estimate the magnitude of such effects.

Ways of Conflict Resolution

The unequal inter-sectoral allocation of water would perhaps be acceptable if water were to be transferred from agriculture to industry without causing loss to agricultural output. This can be expected to happen if water use efficiency in agriculture is raised, through the maintenance and improvement of canals, construction of water courses through which water can be directly taken to the desired crop land and switching over to drip irrigation (wherever possible). The industry, by financing the infrastructure for drip irrigation, would reduce the water requirement for agriculture. However, the industry has to come forward and show its willingness to pay to the farmers for obtaining water over which the farmers have equal rights. The industry's willingness to bear a part of the O&M costs of drip irrigation would also reduce the farmers' hostility. Possibility of all these can be realized when farmers' rights over the Hirakud water is duly recognized by the state and there are strong WUAs. Any decision to change the inter-sectoral water allocation by the government must take these facts into consideration. In the course of time, the cropping pattern can be influenced in favor of less water-intensive crops by creating demand for them, and this demand can be created by agro-based industries that are to be established by either the government or the private sector which plans to share the Hirakud water. If all these processes are set in motion, a win-win situation for agriculture and industry can be achieved in the near future.

≈

Notes

1. Higher Secondary level means class XI–XII level.
2. In 1992, Sambalpur district was divided into Sambalpur, Bargarh, Jharsuguda and Debagarh.
3. See "30,000 Farmers Demand Hirakud Dam Water," *Down to Earth*, December 31, 2007, http://www.downtoearth.org.in/node/7037 (accessed August 9, 2013).
4. See note 3 and "Farmers Protest Water to Industries in Orissa," *OneIndia*, November 6, 2007 www.news.oneindia.in/2007/11/06/farmers-protest-water-to-industries-in-orissa-1194361471.html (accessed May 15, 2008).
5. One million acre foot is equal to 0.12334818375 million hectare meters.
6. A number of sedimentation surveys were conducted in 1979, 1982 and 1986 to assess the rate of sedimentation and storage capacities at different levels of the reservoir (for details, see Mukherjee et al. 2007). These studies reveal that both the dead storage capacity and the live storage capacity of Hirakud reservoir have fallen. In their study through remote sensing techniques, Mukherjee et al. also have estimated the capacity loss to be 24.10 percent from 1957 to 1989 (ibid.). Further, the CWC study (1995) has estimated the reduction of live storage of the reservoir to be 4 MAF (4934 million cubic meters [mcum]).
7. From the reservoir water is taken to Burla power house for electricity generation. Then, the water is released through a power channel. But part of this water is taken to Chipilima power house for power generation. Thereafter, this water is released to the river for irrigation and other uses.
8. The full-fledged power generation took place from 1966.
9. Minors and sub-minors are tertiary canals within the irrigation network.
10. A study by Meher and Baboo (2003) has pointed to illegal cultivation on the acquired land on both sides of the embankment of Bargarh canal which has generated caste conflicts between the encroachers and the land-owning community of villages, such as Mahulpali, Budelpali and Bhabanipali, since such illegal cultivation has reduced the irrigation water available to the land-owning farming community of these villages. However, further study is required to detect the number of such enchroachers and estimate the area encroached.
11. For monthly data on the evaporation of water from the reservoir, see Government of Orissa 2007a, Table 10. However, data on canal-level evaporation is not available.
    12.006 MAF is drawn annually of which 0.004 MAF is drawn during the non-monsoon period (Government of Orissa 2007).
13. Orissa State Pollution Control Board has attempted a use-based classification of Mahanadi water for 2002 at nine different monitoring stations.

According to it, the station Sambalpur downstream is classified D/E. The major causes of worry here are total coliform (TC) and biochemical oxygen demand (BOD). The former means a large number of bacteria present in water, affecting its quality. BOD is the amount of oxygen required by micro-organisms in water to decompose organic waste present in it. Higher the waste, higher will be the BOD, resulting in the shortage of dissolved oxygen for fish and other aquatic organisms. For details, see ADB and DoWR 2010; and Government of Orissa 2004.

14. Cusec is a measure of flow; it means cubic feet per second i.e., 28.317 liters per second.
15. See Government of Orissa 2007b, paragraph 5.4.2.
16. INDAL, Orient Paper Mill and Tata refractory drew 10 cusecs, 24.76 million gallons and 67.4 million gallons of water respectively (Mohanty 2008).
17. Ib and Veden are distributaries of Mahanadi near Hirakud.
18. Riparian water right, originating from English common law, is a system of allocating water among those who own land near its source. These rights are saleable or transferrable other than with the adjoining land, nor is water transferrable out of the watershed.
19. The provisions within SAP were based on the Anderson Memorandum titled "Trade Reforms in India" dated November 30, 1990, submitted to the Government of India by the World Bank. India embarked upon a path of liberalization in the 1980s, whose pace quickened radically after 1985. The net outcome was that India's external debt tripled during this decade of high growth. For more details, see http://www.ieo.org/world-c10-p1.html. The aforementioned website is largely critical of SAP. However, there is also literature supporting the introduction of SAP in India. And the general view has been that it was introduced to improve the efficiency in the different sectors of the economy. The net outcomes have, however, been mixed (see, for example, Ahluwalia 2000, 1994; Nagaraj 1997).
20. Letter No-Irrigation-III HKDW 6/90-40945 dated November 26, 1990, of the Irrigation Department, Government of Orissa. This letter is also mentioned in paragraph 6.2.6, page 33 of Government of Orissa 2007a.
21. One gallon is equal to 0.003785412 cubic meters; so 3191.2 million gallons would be approximately 12.08 million cubic meters (mcum) per year.
22. This was preceded by National Water Policy of 1987. Orissa, too, drafted its (first) water policy in 1994.
23. "CM Announces Rs 200 Crore for Repair of Hirakud Canal System," *Odishadiary*, November 26, 2007, www.orissadiary.com\currentNews.asp?id=4990 (accessed September 17, 2013).
24. Some such instances are presented in Mohanty 2008.

25. It is approximately 9 percent of the live storage as mentioned earlier.
26. Rather, they are often transferred back to the reservoir as waste water causing water pollution.
27. The government's control over water allocation in China has favored urban and industrial claims against agricultural needs. This has happened because at the stage of actual inter-sectoral allocation, the voices of the powerful urban and industrial water users invariably have overridden the claims of rural people. In Philippines, urban users of Manila municipal area and agricultural users both have rights over the Angat Reservoir. But adjustments to shortages are heavily skewed against the interests of the farmers because of the political strength of the Metropolitan workers and Sewerage System in Manila (UNDP 2006).
28. That the return from industrial water use would be higher than the cost incurred in making such provisions for the farmers can prove to be an incentive for the industries to undertake such initiatives.
29. The study of the water quality of Mahanadi river, particularly in its upper reaches, was conducted by University College of Engineering and School of Life Science, both in Burla, from 1983 to 1988 with the support of Department of Environment, Government of Orissa (Government of Orissa 2004).

References

"30,000 Farmers Demand Hirakud Dam Water," *Down to Earth*, December 31, 2007, http://www.downtoearth.org.in/node/7037 (accessed August 9, 2013).

"CM Announces ₹200 Crore for Repair of Hirakud Canal System," *Odishadiary*, November 26, 2007, http://ww.orissadiary.com\currentNews.asp?id=4990 (accessed September 17, 2013).

"Farmers Protest Water to Industries in Orissa," *OneIndia*, November 6, 2007, http://www.news.oneindia.in/2007/11/06/farmers-protest-water-to-industries-in-orissa-1194361471.html (accessed May 15, 2008).

Ahluwalia, M. S. 2000. "Economic Performance of States in the Post Reform Period," *Economic and Political Weekly*, 35(19): 1637–1648.

———. 1994. "Structural Adjustment and Reform in Developing Countries." Paper presented at the conference sponsored by G-24 on the occasion of the 50th anniversary of the Bretton Woods Conference, Cartagenn, Columbia, 18–20 April. Also available online at http://planningcommission.gov.in/aboutus/speech/spemsa/msa013.pdf (accessed September 16, 2013).

Asian Dvelopment Bank (ADB) and Department of Water Resources (DoWR), Government of Orissa. 2010. *TA-7131 (IND)-Technical Assistance for*

Institutional Development of Integrated Water Resource Management in Orissa: An Input Into the Integrated Irrigated Agriculture and Water Management Investment Programme (OIIAWMIP), http://www.indiaenvironmentportal.org.in/files/FINALREPORT-IWRM-ORISSA.pdf (accessed August 16, 2008).

Baboo, B. 2009. "Politics of Water: The Case of Hirakud Dam in Orissa, India," *International Journal of Sociology and Anthropology*, 1(8): 139–44.

D'Souza, R., P. Mukhopadhyay and A. Kothari. 1998. "Re-evaluating Multi-purpose River Valley Project: A Case Study of Hirakud, Ukai and IGNOU," *Economic and Political Weekly*, 33(6): 279–302.

Development Support Centre (DSC). 2003. *Tail-enders and Other Deprived in the Canal Water Distribution*. Final report prepared for Planning Commission, Government of India, http://planningcommission.nic.in/reports/sereport/ser/std_prbirrg.pdf (accessed August 9, 2013).

Gandy, M. 2005. "Learning from Lagos," *New Left Review*, 33 (May/June): 37–52.

Government of Orissa. 2008–09. *Annual Report*, Introduction, p. 9. Department of Water Resources: Government of Orissa.

———. 2007a. *Report of the High Level Committee to Study Various aspects of Water Usage for Hirakud Reservoir*. Bhubaneswar: Government of Orissa.

———. 2007b. *State Water Policy 2007*. Bhubaneswar: DoWR, Government of Orissa.

———. 2004. *Orissa State Water Plan 2004*. Bhubaneswar: DoWR, Government of Orissa.

———. 2003. *State Water Policy 2003*. Bhubaneswar: Department of Water Resources (DoWR), Government of Orissa.

Hussain, I. 2005. *Pro-poor Intervention Strategies in Irrigated Agriculture in Asia*. Final Synthesis Report. Colombo: International Water Management Institute (IWMI).

Meher, Rajkishor and Balgovind Baboo. 2003. "Executive Summary: Tail-enders and Other Deprived in an Irrigation Area: The Case of Orissa in Eastern India," in DSC, *Tail-Enders and Other Deprived in the Canal Water Distribution*, final report prepared for the Planning Commission, Government of India, http://planningcommission.nic.in/reports/sereport/ser/std_prbirrg.pdf (accessed August 9, 2013).

Meinzen-Dick, R. S. and Claudia Ringler. 2006. "Water Reallocation: Challenges, Threats, and Solutions for the Poor," in United Nations Development Programme (UNDP), *Human Development Report 2006: Beyond Scarcity: Power, Poverty and the Global Water Crisis*, p. 238. New York: Palgrave Macmillan.

Mohanty, S. 2008. *Chasira Rekha* (*Line of the Farmer*). Sambalpur, Orissa: Western Orissa Farmers Organization, Integration Samiti.

Molle, F. and J. Berkoff. 2006. "Cities versus Agriculture: Revisiting Inter-sectoral Water Transfer, Potential Gains and Conflicts." Comprehensive

Assessment Research Report 10, Comprehensive Assessment of Water Management in Agriculture, IWMI, Colombo.

Mukherjee, S., V. Veer and S. K. Tyagi. 2007 "Sedimentation Study of Hirakud Reservoir through Remote Sensing Techniques," *Journal of Spatial Hydrology*, 7(1): 122–30.

Nagaraj, R. 1997. "What Has Happened since 1991: Assessment of India's Economic Reforms," *Economic and Political Weekly*, 32(44–45): 2869–79.

Namara, R. E., B. Upadhyay, R. K.Nagar. 2005. "Adoption and Impacts of Microirrigation Technologies: Empirical Results from Selected Localities of Maharashtra and Gujarat States of India." Research Report 93, IWMI, Colombo.

Narayanamoorthy, A. 1997. "Drip Irrigation: A Viable Option for Future Irrigation Development," *Productivity*, 38(3): 504–11.

National Remote Sensing Agency (NRSA). 2004. *Performance Evaluation of Hirakud Project Command Area Using Satellite Remote Sensing Technique.* Hyderabad: DoWR Department, Government of Orissa.

Nayak, A. B. 2006. "An Analysis Using LISS III Data for Estimating Water Demand for Rice Cropping in Parts of Hirakud Command Area, Orissa, India." M. Sc. thesis, Indian Institute of Remote Sensing-National Remote Sensing Agency (IIRS-NRSA), Department of Space, Government of India, Dehradun, India, and International Institute for Geo-information Science and Earth Observation, Enschede, The Netherlands.

Panda, D. K. and S. K. Nayak. 2005. "Dam and Development: A Hirakud experience," *Political Economy Journal of India*, 14(2): 72–77.

Panda, R. K. 2008. "A New Dimension of Water Conflict in Orissa: Industry vs Agriculture," Society for the Study of Peace and Conflict (SSPC) Fostering Ideas, Research and Dialogue, Article 40, SSPC, New Delhi.

———. 2007. "Industry vs Agriculture: The Battle over Water in Hirakud," http://infochangeindia.org/20070108389/Water-Resources/Features/Industry-vs-agriculture-The-battle-over-water-in-Hirakud.html (accessed August 9, 2013).

Pati, B. and Manas Biswal. 2009. "Hirakud Dam: Fifty Mournful Years," *South Asia Network on Dams, Rivers and People*, 7(5–7): 7–11.

Postel, Sandra. 1999. *Pillars of Sand: Can the Irrigation Miracle Last?* New York and London: W. W. Norton & Co.

Rathore, D. S., Anju Choudhary and P. K. Agarwal. 2006. "Assessment of Sedimentation in Hirakud Reservoir Using Digital Remote Sensing Technique," *Journal of the Indian Society of the Remote Sensing*, 34(4): 377–83.

Raul, S. K., S. N. Panda, H. Hollaender and M. Billib. 2008. "Sustainability of Rice-dominated Cropping System in the Hirakud Canal Command, Orissa, India," *Irrigation and Drainage*, 57(1): 93–104.

Sivanappan, R. K. 2002. "Strength and Weaknesses of Growth of Drip Irrigation in India," in *Proceedings of Micro Irrigation for Sustainable Agriculture.* GoI Short-term training June 19–21, WTC, Tamil Nadu Agricultural University, Coimbatore.

Sugunan, V. V. 1995. "Reservoir Fisheries of India." Food and Agriculture Organization (FAO) Fisheries Technical Paper 345, FAO Corporate Document Depository, FAO, Rome. Also available online at http://www.fao.org/docrep/003/v5930e/v5930e00/HTM (accessed August 16, 2008).

United Nations Development Programme (UNDP). 2006. *Human Development Report 2006: Beyond Scarcity: Power, Poverty and the Global Water Crisis.* New York: Palgrave Macmillan.

Villarejo, D. 1997. *Mendota Executive Summary*, http://www.whiteknight.com/Alliance/mendota.htm (accessed August 9, 2013).

10

Sustainable Management and Regional Cooperation for Himalayan Waters

Ramesh Ananda Vaidya and
Madhav Bahadur Karki

The rapid retreat of the Himalayan glaciers may lead to water-related hazards, such as glacier lake outburst floods (GLOFs), and water stress, as a result of the decline in freshwater supplies during the lean season. Thus, there is a need to think and act seriously about promoting cooperation among the countries in the Himalayan region for managing water resources and water-related hazards. According to the Fourth Assessment Report of the United Nations' Inter-governmental Panel on Climate Change (IPCC 2007), the incidence and intensity of floods in the Himalayan region are expected to increase as a result of an increase in extreme precipitation events during the monsoon season and glacial retreat, both, in turn, resulting from global warming. This poses a challenge for reducing the vulnerability of the more than 1.3 billion people living in the major river basins downstream from the Hindu Kush–Himalayan region. The overriding importance of climate change as a driver of environmental change makes it crucial to address disaster-reduction and water-management concerns in a holistic manner at the river basin level. Such an approach is being considered by the IPCC to be an adaptive measure for tackling the impact of climate change.

The economics of water resource management also suggests the need for taking advantage of externalities while planning water management, i.e., production externalities reflected in the upstream–downstream linkages between communities, districts and provinces within national borders, or across international boundaries. While holistic basin-wide water resource management is an approach currently being promoted by water scientists and economists alike, in the Himalayan region the problem lies in the implementation of such a strategy because most of the Himalayan rivers are international rivers and involve transboundary water management.

It has been suggested recently that regional cooperation on water management and water-related hazards can be facilitated by developing a broader perspective on regional economic cooperation that goes beyond the focus on water alone. This perspective would be based on water as a natural resource of central focus, around which cross-border economic exchanges (primarily trade) and development of infrastructure to facilitate them, can take place. This essay looks at these issues under three sections covering: (*a*) conflict and cooperation, (*b*) climate change impact and regional cooperation on water-related hazards, and (*c*) basin-wide regional economic cooperation.

Conflict and Cooperation

This section presents an overview of the fundamental factors behind cross-border water-related conflicts and likely avenues for cooperation, and discusses them using examples drawn from the greater Himalayan region. The basic regional statistics related to the major rivers and river basins are summarized in Table 10.1.

Factors behind cross-border water-related conflicts: Some of the critical indicators of vulnerability to conflict among nations, related to water availability, are the per capita water availability, the level of water withdrawals for annual use in relation to its availability and the extent of dependence on water resources that flow in from the borders. Table 10.2 shows the countrywise trend in per capita water availability in 2000 and 2005. The rate of population growth in these countries is the principal driver for the trend observed in this table.

The critical stress level of water availability, i.e., the level at which users begin to experience shortage of water, has been given

Table 10.1

Principal rivers of the Greater Himalayan Region: Basic statistics

	Area (sq. km)	*Mean Discharge (cubic meter per second [m^3/s])*	*% of Glacier Melt in River Flow*	*Population (1000s)*	*Population Density (no. of persons per sq. km)*	*Water Availability per Person (m^3/year)*
Indus	1,081,718	5,533	44.8	178,483	165	978
Ganges	1,016,124	18,691	9.1	407,466	401	1,447
Brahmaputra	651,335	19,824	12.3	118,543	182	5,274
Irrawaddy	413,710	13,565	Small	32,683	79	13,089
Salween	271,914	1,494	8.8	5,982	22	7,876
Mekong	805,604	11,048	6.6	57,198	71	6,091
Yangtze	1,722,193	34,000	18.5	368,549	214	2,909
Yellow	944,970	1,365	1.3	147,415	156	292
Tarim	1,152,448	146	40.2	8,067	7	571
Total				1,324,386		

Source: Xu et al. 2007.

Table 10.2
Per capita water availability in 2000 and 2005 (m³/person/year)

Country	*River Basin*	*Population (1000s)*	*Per Capita Water Availability* (2000)*	*Per capita Water Availability* (2005)*
Afghanistan	Indus, Tarim	24,926	2,986	2,610
Bangladesh	GBM**	149,664	8,809	8,090
Bhutan	GBM	2,325	45,564	40,860
China	GBM, Indus, Tarim	1,320,892	2,259	2,140
India	GBM, Indus	1,081,229	1,880	1,750
Myanmar	GBM	50,101	21,898	20,870
Nepal	GBM	25,725	9,122	8,170
Pakistan	Indus, Tarim	157, 315	2,961	1,420

Source: WWAP 2006.
Notes: *Water Availability: Total Actual Renewable Water Resources.
**GBM: Ganges–Brahmaputra–Meghna.

as 1700 m^3 per person per year. Some hydrologists have estimated 1,000 m^3 per person per year as the minimum water requirement for a moderately industrialized nation that uses water efficiently (Serageldin 1995). The annual water availability in Pakistan was already below the critical stress level in 2005, and judging from the rate at which it declined between 2000 and 2005, it may soon fall below the minimum level. The data shows that India, China and Afghanistan are also water-starved nations in the region, where the annual water availability is quickly approaching the critical stress level.

Just as population growth can adversely affect the demand side, climate change may have a serious effect on the supply side of water resource management. Table 10.3 shows, countrywise, the ratio of annual water withdrawals (demand) to annual renewable water availability (supply). Among the countries in the region, the level of water demand is about three-fourth of the level of supply in Pakistan. This suggests that there might be a serious water shortage, if the water supply decreases due to adverse consequences of climate change. Normally, the level of demand greater than one-third of the supply is considered risky. On the other end, the ratio is less than 5 percent in Myanmar and Nepal, which is an

Table 10.3
Country-wise ratio of water demand to water supply

Country	*River basin*	*Water Withdrawals as a Percentage of Renewable Supply*
Afghanistan	Indus, Tarim	36
Bangladesh	GBM	7
Bhutan	GBM	NA
China	GBM, Indus, Tarim	NA
India	GBM, Indus	34
Myanmar	GBM	3
Nepal	GBM	5
Pakistan	Indus, Tarim	76

Source: WWAP 2006.

indicator of the vast potential of under-utilized water resources that these countries might be able to tap without harming the riparian nations.

Since the countries in the Himalayan region fall within the three major river basins of the Indus, Ganges–Brahmaputra–Meghna (GBM), and Tarim, the extent to which water resources are shared would be an important indicator of their vulnerability to the competing interests of the water users of these nations. Table 10.4 presents data on the fraction of the total water supply of the countries that originates outside their borders and that flows across their borders to other nations. Both Bangladesh and Pakistan receive more than three-fourth of their surface water supply from across their borders, mainly from India. Furthermore, although only about one-third of water supply to India originates outside its borders, almost three-fourth of the surface water during the dry season in the fertile and densely populated Ganga basin flows from Nepal (WECS 2002). These interrelationships between Bangladesh, India, Nepal and Pakistan in the GBM and Indus basins have the potential to give rise to frictions and tensions over water in future in the region.

Potential Areas for Regional Cooperation on Water

The literature suggests that although cooperation can occur when mutual benefits are possible, existence of potential mutual benefits

Table 10.4
Dependence on imported surface water

Country	*River Basin*	*Percentage of Total River Flow Originating Outside the National Border*
Afghanistan	Indus, Tarim	15
Bangladesh	GBM	91
Bhutan	GBM	0.4
China	GBM, Indus, Tarim	1
India	GBM, Indus	34
Myanmar	GBM	16
Nepal	GBM	6
Pakistan	Indus, Tarim	76

Source: WWAP 2006.

alone is not sufficient for cooperation to take place. There is a need to ask three questions in this regard. Are there truly potential mutual benefits or is it a situation where only one party can benefit at the cost of the other? In case of the latter, can the situation be redefined in order to be transformed into to one of potential mutual benefits? What are the impediments to actually achieving mutual benefits? (Crow and Singh 2000). Cooperation between India and Pakistan over the Indus river basin is considered a good example of the situation being redefined to transform into one of potential mutual benefits by enlarging the size of the pie rather than just dividing it.

The principal potential benefits of cooperation in water resource management are: (*a*) sharing information for flood forecasting and early warning, (*b*) storing water in upstream river basins for flood moderation in downstream areas, (*c*) storing water resources in order to increase the flow during the dry seasons, (*d*) storing water for inland water transport, (*e*) harnessing water resources to generate hydroelectricity and (*f*) managing watersheds to help increase the quality and quantity of water available for irrigation and drinking water by downstream users.

The type of exchange of benefits between countries may be: (*a*) bilateral barter, which is subject to the need to find a "double coincidence of wants," or (*b*) a financial transaction based on the payment of a mutually agreed upon monetary value for the environmental services delivered. Table 10.5 lists what the

Table 10.5
A simple framework to study potential water-related international transactions between Bhutan, India and Nepal

Potential Parties	*Good or Service*	*Type of Exchange Anticipated*
Bhutan to India	Supply of hydroelectric power#	Monetized
Bhutan to India	Supply of water storage benefits*	Barter exchange
India to Bhutan	Navigation and transit#	Barter exchange
India to Bhutan	Provision of finance and engineering for construction#	Partly monetized
Nepal to India	Supply of hydroelectric power*	Monetized
Nepal to India	Supply of water storage benefits*	Barter exchange
India to Nepal	Navigation and transit#,*	Barter exchange
India to Nepal	Provision of finance for construction**	Monetized
India to Nepal	Provision of engineering expertise*	Probably monetized

Source: Crow and Singh 2000.
Notes: #: occurring to some extent.
*: discussed.
**: suggested.

governments of Bhutan, India and Nepal have sought from each other to benefit mutually from the development of water resources in the GBM basin.

A Simple Example of Cooperation

The opportunities for cross-border cooperation between India and its neighbors, viz., Bhutan and Nepal, on hydroelectricity were enhanced by two major developments in India related to the power sector, one physical and the other institutional. At the physical level, hydroelectric power-grid interconnections in India evolved from the local level in the 1950s to the provincial level in the 1960s, and then on to the regional level in the 1970s and finally to the national level in the 1990s. For the development of transmission and power-grid interconnections, the Power Grid Corporation of India Limited (PGCIL) was given responsibility for: (*a*) the development of a national grid by connecting the five regional grids, (*b*) establishing the national load dispatch center, and (*c*) modernizing the regional and provincial load dispatch centres.

At the institutional level, India established trading companies, such as the government-supported Power Trading Corporation (PTC), to promote the power market. As these physical and institutional set-ups achieved operational success within its borders, they started exploring cross-border sources of power for interconnections and trading. From India's perspective, as the PTC sees it, the rationale for long-term cooperation in energy are: (*a*) to take advantage of the potential for economies of scale as a result of large cross-border projects set up primarily to address opportunities provided by India's shortage situation, and (*b*) to use cross-border renewable energy sources for environmental and conservation benefits, among others. They have identified the major constraints to cross-border power trading: (*a*) transmission infrastructure and wheeling facilities within the national boundaries, (*b*) inter-grid synchronization for cross-border interconnections, and (*c*) electricity pricing.

Worldwide, cross-border grid interconnections are on the rise. Interconnections already exist in north America, Europe and southern Africa, including the Nord Pool (Denmark, Finland, Norway and Sweden) and the South African Power Pool (12 countries). As per our understanding, India intends to use its recent domestic experience in regional grid interconnections for expanding cross-border grid interconnections. India's existing major cross-border interconnections are with Bhutan: Chukha hydroelectric project (336 MW) connected at Birpara in India, and Tala hydroelectric project (1,020 MW), at Siliguri in India.

For electricity pricing, the power trading companies act as "market makers," negotiating prices separately with the producers (generators) and distributors and thus taking the "market risk" which they tend to diversify by dealing with a large number of buyers and sellers, in what is a suppliers' market as of now. The trading is largely short-term, i.e., for less than a year, although efforts are being made to increase the long-term ones, to achieve a 70–30 short-term–long-term mix.

However, such cooperation in hydroelectric power can take place only after the necessary legal provisions are agreed upon by the two countries. The India–Nepal Power Trade Agreement, signed on June 5, 1997 but yet to be ratified by the Nepalese legis-lature, is an umbrella agreement for power trading between the two nations (GoN 1997). It goes a step ahead of the agreements between Bhutan and India that are dedicated to specific hydroelectric projects. The India–Nepal agreement provides unlimited marker access for power trading. Any party in India or Nepal may enter into a power trade agreement: governmental, semi-governmental, or private enterprise (Article 1). It also makes a provision for the market mechanism to decide on the price and quantum of electricity to be delivered at a mutually agreed-upon destination, without any form of government intervention (Article 2). Once it is fully operationalized, it may be the best practice example of cooperation in power trade between the riparian countries. To realize such opportunities for cooperation in hydropower generation, however, it may be increasingly necessary to take adaptation measures to face potential GLOFs, transboundary floods and droughts in the region.

Climate Change Impacts and Regional Cooperation

Climate change has introduced a new dimension to the potential benefits of cooperation in the context of upstream–downstream linkages. Temperature changes in the Himalayas have been much higher than the global average. In Nepal, an increase by 0.6^0 C has been recorded every 10 years between 1977 and 1999 (Shrestha et al. 1999). Warming in Tibet has also been progressively higher with elevation (Liu and Chen 2000). The Himalayan glaciers, especially in the eastern and central regions, have been shrinking at an accelerated rate in recent decades, although this drastic reduction in ice cover has not been observed in the north-western Himalayas, Karakorum, Hindu-Kush or Pamirs (Xu et al. 2007). However, these observations have to be evaluated carefully as they have been largely based on limited case studies.

Climate change may lead to an increase in the frequency and magnitude of GLOFs and flood-related disasters. Scientists have estimated that if the present trend of glacier melting continues, most valley glacier trunks and small glaciers will disappear by 2050 (Dyurgerov and Meier 2005). As glaciers retreat, glacial lakes are often formed especially at altitudes of over 4,500 m. Subsequently, GLOFs may occur, as the amounts of melt-water in these lakes increase and the moraine deposits are washed away. There have been at least 34 GLOFs in Bhutan, China and Nepal in the last 25–30 years, and 204 potentially dangerous lakes have been identified in the Hindu-Kush–Himalayan region (Ives et al. 2010). In addition, high relief, steep slopes, complex geological structures with active tectonic processes and continued seismic activities, and a climate characterized by great seasonality in rainfall, all combine to make water-induced disasters a common phenomenon in the Himalayan region (Eriksson et al. 2009). Floods and droughts are likely to increase further, both due to a decline in the glaciated area of the basins in the Hindu-Kush–Himalayan region causing reduced hydrological modulation and due to an increase in extreme precipitation events.

A Simple Example of Regional Impact of Potential GLOFs

To demonstrate the growing need for regional cooperation in the context of global warming, an example of upstream–downstream linkages related to potential GLOFs and hydropower stations in the Himalayan region is presented here for the Dudh Kosi sub-basin in the GBM's Kosi basin (54 percent of whose catchment area falls in China [Tibet] and 46 percent in Nepal).

The Dudh Kosi sub-basin is home to about 36 valley glaciers. These glaciers have retreated at rates ranging between 10 and 74 m per year. For example, the rate of retreat for the Imja glacier near Mt Everest, the one retreating fastest, has increased from 41 m per year in 1962–2001 to 74 m per year in 2001–2006 (Bajracharya et al. 2007). This sub-basin contains 12 "potentially dangerous" glacier lakes, all moraine dammed, including Imja and Dig Tsho. The basin has experienced GLOFs in 1977, 1985, 1998 and 2001, including the Dig Tsho GLOF. The Dig Tsho GLOF on August 4, 1985 caused a 10–15 m high surge of water and debris to flood down the Dudh Kosi river in Arun basin river for 90 km. At its peak, the discharge was 5,613 m^3 per second, 2–4 times the magnitude of maximum monsoon flood levels (Shrestha et al. 2006). The flood began in early afternoon and lasted for about six hours. It resulted in the complete destruction of the almost-completed hydropower project at Thame near Namche in the Everest region, built at an estimated cost of US$ 1 million. To minimize the adverse impact of such events, it is necessary to: (*a*) monitor and assess the status of glacier lakes, (*b*) install early warning systems, (*c*) implement mitigation measures, and (*d*) develop estimates of flow-regime changes in different catchments under various likely climate change scenarios, in order to develop a scientific basis for cost-effective adaptation measures (Bajracharya et al. 2007).

It is clear from detail GIS mapping of the Kosi Basin that GLOFs in China (Tibet Autonomous Region) and Nepal can have serious consequences for the existing and planned hydropower stations in Nepal (Map 10.1).

The location of the cross-border potentially dangerous glacial lakes in the catchment of Arun river, on which the Upper Arun (335 MW), Arun III (402 MW) and Lower Arun (308 MW)

Map 10.1
Potentially dangerous glacier lakes and hydropower stations in the Kosi basin

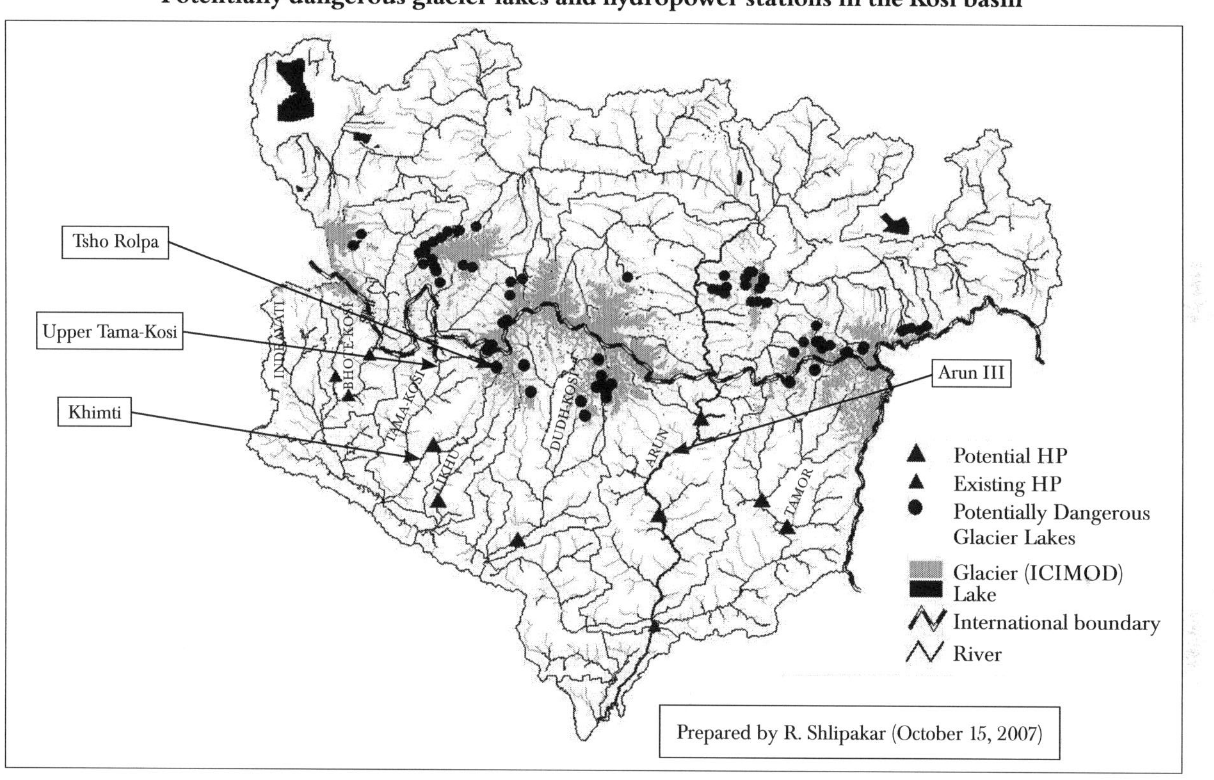

Source: Bajracharya et al. 2007. Map not to scale.

hydroelectric power stations are planned to be built, is noteworthy. The location of such lakes in the catchment of Dudh-Kosi River, on which a storage hydropower project (300 MW) is being planned, also deserves special attention. Further to the west, the Tsho Rolpa glacial lake is situated in the catchment that can affect the tailrace part of the existing Khimti-Khola hydropower station (60 MW). This is also one of the cases in which mitigation measures have been carried out and early warning systems established by the government, with donor support. Further west of the Dudh-Kosi river is the Tama-Kosi river on whose catchment several potentially dangerous glacier lakes have been identified across the border in China (Tibet). These examples demonstrate that regional cooperation in managing GLOF risks is essential for reducing project risk and, in turn, for raising funds at a reasonable cost of capital for generation of hydroelectricity in Nepal which could then be sold in the Indian market at competitive prices.

Specific areas of action for regional cooperation are:

(*a*) The Himalayan region is considered a data gap or "white spot" region in the global climate map of the Fourth Assessment Report of the IPCC due to the paucity of data on the region's hydrology and meteorology. There is, therefore, a need for regional cooperation among the countries in the Himalayan region to gather and share information for assessing and monitoring climate change and its consequences for water resource management.

(*b*) National Adaptation Plans of Action (NAPAs) have been prepared by the governments of the Himalayan region on the initiative of the UN Framework Convention on Climate Change (UNFCCC). These NAPAs suggest plans of action to utilize satellite-based techniques and field-based techniques simultaneously for monitoring glacier retreat and assessing the potential impact of GLOFs on downstream areas.

(*c*) The principle of integrated water resource management (IWRM) at the basin level has already been accepted by scientists and policy-makers alike. It is considered by the IPCC to be an adaptive measure for tackling the impact of climate change. The concern currently is about how to implement this approach by incorporating water governance within the framework of national governance. As the national governments of the Himalayan region make preparations to implement IWRM, it is necessary to consider regional cooperation for adaptation to the events occurring in the catchments that lie across the borders. For example, as described earlier, GLOFs in China can affect hydropower stations in Nepal.

Further, the concept of IWRM at the basin level follows an ecosystem-based framework. Water is considered a lifeline or blood supply line of the entire ecosystem, climate and environment as a whole. The inherent nature of the geo-morphological make-up of the catchments of the river and the connecting sub-rivers determines the potential of natural siltation. The human-induced changes and infrastructure development, combined with the regional and global climate change impacts, indicate the aggravated nature of hazards that can trigger water-induced disasters. The concept of IWRM at the basin level can promote an integrated management of land, water and plant resources involving people within the catchment areas. Such an approach has already been practiced to some extent, in the Mekong river basin.

Basin-wide Regional Economic Cooperation

The state of international cooperation on water is limited by the fact that governments tend to negotiate agreements on benefit-sharing basis in very specific areas of cooperation, such as hydroelectric power trade, without approaching it in a holistic manner. While it is a step in the right direction for developing physical and institutional mechanisms for cooperation on water, the limitation of such cooperation is that it does not clearly address the factors leading to cross-border conflicts on water, such as the declining per capita water availability due to increasing demand in both domestic and industrial sectors, depleting supply and increasing population growth in urban centers – all increasing the gap between supply and demand.

Treaties on international rivers do not seem to reflect the concerns that the scientific and economic principles discussed earlier suggest. In an analysis of 145 international treaties related to water, 39 percent included benefit-sharing in hydroelectricity. In only about a third of the cases was the quantum of water allocation considered, with flood control accounting for 9 percent (UNDP 2006).

There is also a preference among riparian nations for bilateral negotiations, even in river basins shared by more than two countries. Out of the 263 international river basins, 106 have water institutions and two-thirds of these have three or more riparian nations. However, less than a fifth of the accompanying agreements are multilateral, showing the preference of riparian countries for bilateral agreements (UNDP 2006). Furthermore, third-party mediation is also discouraged, even when there are asymmetries of power between the countries involved. Here, the main issues appear to be those related to property rights and externalities. The fact that international river water is a common property of all the basin nations with a high "free rider" tendency by the lower riparian countries has limited exclusion potential, especially for economically weaker parties.

Many international river basins have shown interest in following an approach of cooperation at the basin level. Such cooperation may be in the form of: (*a*) coordination of activities, such as information-sharing, (*b*) collaboration among the nations, such as in developing adaptable plans, or (*c*) common action by the riparian countries, such as joint/collective development of infrastructural facilities. The following statement by Nepal at the UN General Assembly meeting in September 1998 probably sums up the interests of both upstream and downstream nations in the Himalayan region:

> Mr. President: On our march towards a democratic and just society, we face many challenges, such as pervasive poverty, mass illiteracy, environmental degradation, population explosion and, above all, gender inequality. We believe that many problems related to economic development can be more effectively tackled through

> regional or sub-regional cooperation among nations. Tremendous opportunities are available for sub-regional cooperation in our part of the world among the countries in the Ganga-Brahmaputra-Meghna basin. These opportunities include water resources development, flood control, energy supply, forestry management and environmental protection, among others. Development efforts in water resources, for example, would help irrigate the fertile fields in the plains of India, improve the waterways so vital for the transportation sector of Bangladesh, and generate hydropower in Nepal to meet the energy needs of the region as a whole. Such a development strategy may be the key to future prosperity in the region (UN General Assembly 1998).

Interestingly enough, such an environment for cooperation may be feasible only if we can also consider enlarging the basket of benefits resulting from cooperation, indirect economic benefits "beyond water," although water will be the main focus. Table 10.6 presents a simple framework on the basket of benefits resulting from cooperation among nations in the basin. This framework has adopted the classification of benefits into four types: political, environmental, direct economic and indirect economic. All four are defined and illustrated in Table 10.6 (Sadoff and Grey 2002). The table shows how the domain of treaties may expand from those based on the conventional practice of water-sharing to (*a*) the optimum utilization of water resources in the basin through IWRM, and finally to (*b*) regional economic cooperation through the integration of regional infrastructure, trade and markets.

The success of all these treaties, however, will depend on the strength of the institutional mechanisms the riparian nations in the basin adopt for water governance. It appears to us that, in the changing context of regional integration for trade and investment among nations, institutional mechanisms for regional economic cooperation with a focus on water may have the highest probability of success.

The Greater Mekong Sub-basin (GMS) program in the Mekong basin may be a good example of an institutional mechanism for regional economic integration. The basin is shared by the six nations, viz., Cambodia, China, Laos, Myanmar, Thailand and Vietnam. Along the lines of the classification in the framework depicted in Table 10.6, while there have been major disputes in

Table 10.6
A simple framework on the benefits of cooperation

Taxonomy of Benefits	*Definitions*	*Examples*	*Domain of Treaties*	*Probability of Success*
Political benefits	Reducing costs because of the river	Policy shift from conflicts to cooperation and development	Conventional practice of water-sharing	Low to medium
Environmental benefits	Increasing benefits *to* the river	Improved water quality, river flow characteristics, soil conservation, biodiversity	Based on the principles of IWRM	Medium, e.g., Mekong River Basin Agreement
Direct economic benefits	Increasing benefits *from* the river	Improved water resource management for drinking, irrigation, navigation and hydropower generation; for the conservation of freshwater ecosystems; and for risk-management of water-related disasters	Based on the principles of IWRM	Medium, e.g., Mekong River Basin Agreement
Indirect economic benefits	Increasing benefits *beyond* the river	Integration of regional infrastructure, markets and trade	Based on the principles of regional economic cooperation	Medium to high, e.g., GMS program

Source: Adapted from Sadoff and Grey 2002.

the region "because of the Mekong," significant benefits have also been derived "from the Mekong" through the lower basin's cooperative management. Furthermore, sharing these benefits has not only been an important stabilizing factor in the lower basin, but has also brought substantial economic benefits "beyond the river," both directly and indirectly. These benefits "beyond the river" include the hydroelectric power trade between Laos and Thailand, even during the periods of conflict and natural gas purchase by Thailand from Myanmar, creating ties that bind the countries in a web of mutual dependency.

An Example of Basin Wide Cooperation on the Mekong River

The Mekong River Basin Agreement of 1995 is an example of a treaty that emphasizes on IWRM. The purpose of the agreement is to support the sustainable development and management of basin's water and related resources; addresses institutional, financial and management issues; and thereby enhance coordination between the member countries. Prior to its signing in 1995, the Mekong Regional Commission's (MRC's) forerunner, the Interim Mekong Committee, examined the legal and institutional structures in other river basins (Mekong Secretariat 1994). The Mekong River Basin Agreement benefitted from inputs from, among others, the Murray-Darling Basin Commission in Australia.

Four of the Mekong basin countries, Cambodia, Laos, Thailand, and Vietnam, have cooperated in the sustainable development and management of the Mekong river basin water resources. The MRC consists of three permanent bodies:

(*a*) The Ministerial Council is the senior body which is made up of one member from each participating country at the ministerial or cabinet level, who is authorized to make policy decisions on behalf of the government (MRC 1995, Article 15).
(*b*) The Joint Committee consists of country representatives from relevant government ministries at the Head-of-Department level (MRC 1995, Article 23).

(*c*) The MRC Secretariat provides technical and administrative services to the Council and the Joint Committee, according to the directions of the Joint Committee (MRC 1995, Article 28).

In addition, as outlined in the agreement, there are National Mekong Committees in each of the participating countries which act as liaison bodies between the MRC Secretariat and the national organizations. Based on the experiences gained in sub-Saharan Africa, Southeast Asia and South Asia (Salman and Uprety 2002; Chenoweth et al. 2001; Gould and Zobrist 1989), the constraints to the management of such regional basin organizations have been identified as: constrained autonomy; weak institutional capacity; insufficient financing; inability of institutions to enforce agreements; lack of expertise for technical, social and environmental analyses required to formulate regional water resource development plans; unavailability of financial resources to provide appropriate levels of support for planning studies and to manage and operate the river basin organizations; lack of institutional capacity for as well as commitment to project implementation due to the pressure of vested interest groups as opposed to regional development planning; and limited authority vested by the national governments in the regional basin organizations to implement policies and programs within its mandate.

The absence of China and Myanmar in the MRC has prevented full-scale basin-wide management. Nevertheless, they have joined the MRC countries in the GMS program, which has been envisaged from the perspective of regional economic cooperation and promotion of trade and investment in the region. The Asian Development Bank (ADB), through the GMS program, has supported regional cooperation for strengthening cross-border physical connectivity. The key activities of the GMS include development of economic corridors, focus on the improvement of road access through investments; institutional and policy changes for trade facilitation; and transit policy harmonization to reduce logistics costs across the sub-region. Five economic corridors have been identified and several investments are under way in these corridors for improved road access, while feasibility studies are addressing prospective improvements in railway networks. In addition to hard infrastructure facilities, ADB has also focused on cooperation through trade and transit harmonization (Kuroda 2006).

All six nations in the GMS program have agreed to promote trade and investment in the region. Such a perspective of regional economic cooperation goes beyond the focus on water alone, but would be based on water as a natural resource of central focus, around which regional economic exchange, primarily trade and infrastructure development takes place.

Lessons Learned for Regional Cooperation

In the changing context of the regional cooperation scenario brought about by climate change and economic globalization and their consequences for water stress and water-related hazards, the importance of wise management of transboundary rivers cannot be over-emphasized. The international protocols and conventions, as well as best practices and experiences discussed in this essay, provide good models to conceptualize, design and promote future cross-border cooperation. Based on the analysis in this essay, the following policy implications and conclusions can be drawn. First, one needs a comprehensive ecosystem management framework for the development and management of water resources in the Himalayas. Second, the impact of climate change could be a major problem in future water management. Therefore, we need to focus holistically on river basin management and act expeditiously. This may mean working across boundaries. In the context of upstream–downstream linkages, it is necessary to consider the benefits of transboundary cooperation for coping with the events occurring in the catchment areas that lie across the borders. GLOFs in China, for instance, can affect hydropower stations in Nepal. There is also a need for transboundary cooperation among the countries in the Himalayan region to gather and share information related to water-induced hazards. To this end, the mechanism for sharing data through a regional inter-governmental institution needs to be strengthened. Third, regional cooperation in water should be made part of the total regional economic cooperation and infrastructure development strategy. The establishment of power grid networks and the proli-feration of power trading companies in India have helped develop the physical and institutional mechanisms necessary for trans-boundary trade in electricity. The imperatives of globalization and climate change are also strongly driving countries to embark on regional cooperation encompassing trade, commerce and other economic exchanges along with regional sustainable development strategies. The perspective of regional economic cooperation may help to expedite the implementation of institutional mechanisms for regional cooperation on water and water-related hazards for which good potential exists in the Himalayan region.

≈

References

Bajracharya, S. R., P. K. Mool and B. R. Shrestha. 2007. *Impact of Climate Change on Himalayan Glaciers and Glacial Lakes: Case Studies on GLOF and associated Hazards in Nepal and Bhutan*. Kathmandu: International Centre for Integrated Mountain Development (ICIMOD).

Chenoweth, J. L., H. M. Malano and J. F. Bird. 2001. "Integrated River Basin Management in the Multi-jurisdictional River Basins: The Case of the Mekong River Basin," *International Journal of Water Resources Development*, 17(3): 365–77.

Crow, B. and N. Singh. 2000. "Impediments and Innovation in International Rivers: The Waters of South Asia," *World Development*, 28(11): 1907–1925.

Dyurgerov, M. D. and M. F. Meier. 2005. *Glaciers and Changing Earth System: A 2004 snapshot*. Boulder, CO: Institute of Arctic and Alpine Research, University of Colorado.

Eriksson, M., Xu, J., A. B. Shrestha, R. A. Vaidya, S. Nepal and K. Sandstrom. 2009. *The Changing Himalayas: Impact of Climate Change on Water Resources and Livelihoods in the Greater Himalayas*. Kathmandu: ICIMOD.

Gould, M. S. and F. A. Zobrist. 1989. "An Overview of Water Resources Planning in West Africa," *World Development*, 17(11): 1717–1722.

Intergovernmental Panel on Climate Change (IPCC). 2007. *Climate Change 2007: The Physical Science Basis*. Contribution of Working Group I to the Fourth Assessment Report of the IPCC. Cambridge: Cambridge University Press.

Ives, J. D., R. B. Shrestha and P. K. Mool. 2010. *Formation of Glacial Lakes in the Hindu Kush-Himalayas and GLOF Risk Assessment*. Kathmandu: ICIMOD.

Kuroda, H. 2006. "Infrastructure and Regional Cooperation." The Asian Development Bank's Keynote Paper presented at the Annual Bank Conference on Development Economics, organized by the World Bank, May 29–30, Tokyo, Japan.

Liu, X. and B. Chen. 2000. "Climatic Warming in the Tibetan Plateau during Recent Decades," *International Journal of Climatology*, 20: 1729–42.

Mekong Regional Commission (MRC). 1995. *Agreement on the Cooperation for the Sustainable Development of the Mekong River Basin*. Bangkok: MRC.

Mekong Secretariat. 1994. *Preparatory Organizational and Legal Studies (Basinwide): Final Report*. Bangkok: Mekong Secretariat.

Government of Nepal (GoN). 1997. *The India-Nepal Power Trade Agreement*. Kathmandu: Ministry of Water Resources (MoWR), GoN.

———. 2002. *Water Resources Strategy: Nepal*. Kathmandu: Water and Energy Commission Secretariat (WECS), GoN.

Sadoff, C. W. and D. Grey. 2002. "Beyond the River: The Benefits of Cooperation on International Rivers," *Water Policy*, 4: 389–403.

Salman, S. and K. Uprety. 2002. *Conflict and Cooperation on South Asia's International Rivers*. Washington DC: The World Bank.

Serageldin, I. 1995. *Toward Sustainable Management of Water Resources*. Washington DC: The World Bank.

Shrestha, A. B., C. P. Wake, P. A. Mayewski and J. E. Dibb. 1999. "Maximum Temperature Trends in the Himalaya and Its Vicinity: An Analysis Based on Temperature Records from Nepal for the Period 1971–94," *Journal of Climate*, 12: 2775–87.

Shrestha, A. B., B. Bajracharya and L. Rajbhandari. 2006. "Glacier Lake Outburst Flood Hazard Mapping of Sagarmatha National Park," in K. P. Sharma and R. Jha (eds), *Proceedings of the Workshop on Flood Forecasting Management in Mountainous Areas*, pp. 228–42. Kathmandu: Department of Hydrology and Meteorology, GoN.

United Nations Development Programme (UNDP). 2006. *Human Development Report*. New York: Oxford University Press.

United Nations General Assembly. 1998. *Statement by Madame Shailaja Acharya, Deputy Prime Minister of Nepal and the Head of the Nepalese Delegation to the General Assembly Meeting*. New York: UN.

Water and Energy Commission Secretariat (WECS). 2002. Water Resources Strategy–Nepal, p. 18. Kathmandu: Nepal.

United Nations Legislative Series. n.d. "The Indus Waters Treaty 1960," in S. N. Bastola (1994), *Water Resources Development of the Mighty Himalayan Rivers*. Kathmandu: Lalita Printers.

World Water Assessment Programme (WWAP). 2006. *Water: A Shared Responsibility*. The United Nations World Water Development Report 2. Paris: WWAP.

Xu, J., A. Shrestha, R. Vaidya, M. Eriksson and K. Hewitt. 2007. *The Melting Himalayas: Regional Challenges and Local Impacts of Climate Change on Mountain Ecosystems and Livelihoods*. Kathmandu: ICIMOD.

PART V

Water in Changing Contexts

11

Growing City, Diminishing Water Access

Urbanization and Peri-urban Water Use in Gurgaon and Faridabad, India

Vishal Narain

"Peri-urban" is a loosely used term denoting contradictory processes and environments (Iaquinta and Drescher 2000). It is interpreted in different ways by scholars, researchers, academics and practitioners. Broadly speaking, however, the word "peri-urban" is used to denote a place, concept or a process (Narain 2009; Narain and Nischal 2007). This chapter seeks to describe the processes through which water use, access and management practices change in peri-urban settlements. The study is based on two villages, each in Gurgaon and Faridabad, two of the fastest growing districts in the north-western Indian state of Haryana, bordering Delhi.

The basic research question that this chapter seeks to address is: how are processes of urbanization affecting water use, access and management practices of peri-urban residents? The study adopts a qualitative research design, relying on an ethnographic approach: a mix of semi-structured interviews, meetings with key informants, direct observation and focus group discussions (FGDs).

The research uses the case study method, which is "an empirical enquiry investigating a contemporary phenomenon within its real-life context, when the boundaries between phenomenon and context are not clearly evident and where multiple sources of evidence are used" (Yin 1984). The case study is used to present a picture of a particular social phenomenon. As against a sample-based study that seeks to generalize the conclusions drawn from a small sample for a larger population – a phenomenon called "statistical generalization" in research parlance – a case study approach seeks to further "analytic generalization," wherein the effort is to generalize an issue or a theory. This implies that the case studies presented in this essay are not of a "general" nature such that the issues described are always to be found in all peri-urban settings – that is not a claim that the essay seeks to make; in fact, given the nature of peri-urban settings such generalizations are not only difficult to make, but also dangerous. The essay seeks to contribute, instead, to an understanding of the diversity of ways in which urbanization affects water use, access and management practices of peri-urban residents. At the same time, some effort has also been made to draw on similar observations made in other peri-urban settings to contextualize the findings of this study better.

The Context of Research: Delhi's Growing Population and Peri-urban Spillover into Gurgaon and Faridabad

The Delhi UA (Urban Agglomeration) has grown by over 4 percent annually in every decade since 1931 (Kundu 2008). This growth has taken place both within and outside the urbanizable limits (ibid.). It is, however, the peripheral areas of the city that have absorbed the majority of the immigrants. This trend, along with a real estate boom and development of major transport corridors, has led to the emergence of a PUI (Peri-urban Interface) in all directions around Delhi.

This growth is particularly pronounced in Gurgaon. The growth potential of Gurgaon city, the headquarters of the Gurgaon district, was quickly realized soon after the Indian economy embarked upon a process of economic reforms and liberalization in the 1990s. There have been several reasons for this growth (Narain 2007). The first of these is Guragaon's proximity to Delhi, the national capital located about 32 km away, in general and the international airport, located about 12 km away, in particular. Further, the state government of Haryana took several policy initiatives to invite industries in Gurgaon following the phase of economic reforms and liberalization initiated in 1991. The most recent such initiative has been the setting up of SEZs (Special Economic Zones). Gurgaon has also seen a real estate boom since the 1990s, and the landscape of the new city is dominated by skyscrapers housing the offices of corporate giants, modern shopping malls and residential facilities. Gurgaon, therefore, has emerged as one of India's major outsourcing hubs, housing major multinationals.

Following closely on the heels of Gurgaon is Faridabad, located, too, in the state of Haryana (Narain and Nischal 2007). Faridabad is about 25 km from Delhi. It is the most densely populated district in the state. With a share of about 5 percent of the total land of the state of Haryana, it accommodates 10 percent of its population. The population density, as per the 2001 population census, is 1,020 persons per sq. km, as against 372 for Haryana as a whole.

The growth of Faridabad has been fueled particularly by a well-connected network of roads and adequate supply of electricity. The Delhi–Mathura National Highway No. 2 (Shershah Suri Marg) passes through the center of the district. A broad-gauge railway line of the Central Railways also passes through the district; most of the trains going to south and west of India pass through the district as well. Faridabad is also a major industrial hub of north India.

However, the phenomenally rapid growth of these cities has put tremendous pressure on their fragile infrastructure and natural resources. The lack of planning to accommodate the new settlers has begun to show its effects, particularly in Gurgaon. Even as residential areas with modern facilities are built, Gurgaon has been plagued by poor infrastructure, especially badly maintained roads, erratic power supply and a growing pressure on its water resources. According to the CGWB (Central Ground Water Board), 70 percent of Gurgaon's water needs are met through groundwater and the water table is dropping at the annual rate of about 1 m. Officially, Gurgaon has been declared a "dark zone" by the CGWB in terms of groundwater over-exploitation. However, this has not stopped the government from maintaining its current pace of urbanization marked by the construction of huge residential complexes and malls. Groundwater is the only source of irrigation for large parts of the district. Where the groundwater is saline, it becomes a major constraint to agriculture, except when this can be overcome through sewage-based irrigation, as will be seen in one of the case studies documented in this chapter.

Faridabad faces a similar stress on its water resources. Isolated groundwater mounds and troughs in different parts of the district are known to have been created because of heavy pumping in the city area (GoI 2007). In general, the water table level has declined all over the district; this trend was particularly conspicuous during the 1980s, when the growth of the city had started picking up. From June 1983 to 1993, a decline in water table level from 1 to 6 m was observed in different parts of the district and was more pronounced in the southern blocks. Besides, drying up of tube wells in the eastern parts of Faridabad and Ballabhgarh blocks is also indicative of the stress on the groundwater resource.

In both the districts, the official response of state agencies to the emerging problems has been supply augmentation. The Union Ministry of Urban Development has approved a project to extend water supply to Faridabad at an estimated cost of INR 490 million.[1] At present, the water supply network in the city is dependent on groundwater resource that caters to about 35 per cent of the city. There are 240 tubewells in the city, of which 43 are unfit for consumption. In a similar vein, keeping in mind the rapidly growing demand for water in Gurgaon, Haryana Urban Development Authority (HUDA) plans to construct a second water treatment plant with a capacity of 66 MGD (Million Gallons a day) at a cost of INR 1.38 billion.[2] HUDA has recently completed the construction of a water treatment plant with a capacity of 20 MGD at a cost of INR 140 million at Basai, one of the four villages which this research is based on and where two water treatment plants of 20 MGD each have already been functioning.

Research Sites

This chapter explores the implications of the aforedescribed trends and processes of water use and management in two peri-urban villages in each of the two districts. The following are the brief profiles of the four villages chosen for this study.

Shahpur Khurd

Shahpur Khurd is a village in the Ballabhgarh block of Faridabad district. It has a population of about 1,000 people. The village is inhabited by three main social groups; *Jaats*, *Harijans*, and *Kolis*. The *Jaats* are socially and economically the most powerful of the three groups. They constitute 70 percent of the village population, the *Harijans* represent 25 percent and the *Kolis* are in a minority. The *kharif* (monsoon season) crops are pearl-millet and sorghum, while the *rabi* (winter season) crops are wheat, *burseem* (a fodder crop) and mustard. This cropping pattern prevails predominantly on account of the limited availability of water for irrigation; the groundwater is saline and only about 10 per cent of the cultivated land is served by canals. This situation does not allow farmers – except a small minority – to cultivate paddy or sugarcane. Borewells are the predominant source of irrigation and diesel-powered borewells are common. Almost all farmers own private borewells. Groundwater is also sold at the rate of INR 40 per hour of water pumped.

Karnera

Karnera also is located in the Ballabhgarh Block of Faridabad district. It has a population of 2,000 people. The main crops grown are wheat, potato and *burseem* in the *rabi* season and pearl-millet, paddy, sorghum and some vegetables in the *kharif* season. Inadequate water availability was cited by the residents as the major factor restricting sugarcane cultivation. Another factor was the distance of the nearest sugar mill which is located at Palwal, about 30 km away. There is a dairy in the village that procures milk from the farmers; the adjoining town of Ballabhgarh also provides a market for the dairy produce. The net cultivated area is about 400 hectares, all of which have been brought under irrigation. Both surface water and groundwater are used for irrigation. Another source of irrigation is the Gurgaon Canal Sewerage Canal originating from Delhi.

Basai

Basai lies in the Gurgaon district of Haryana, about 3 km from the Gurgaon city. It is located adjacent to sectors 9 and 10 of Gurgaon, two of the city's major residential sectors that have been developed by acquiring land from it. Basai has also provided land for the water treatment plant of HUDA that supplies drinking water to much of the Gurgaon city. The village is dominated by the *Jaat*s numerically and in terms of land ownership. The village also has a substantial immigrant population from Rajasthan and Bihar constituting the laborforce for agriculture and construction. The main crops grown are wheat, sorghum, pearl-millet and fodder crops. In addition, some farmers are able to cultivate paddy that is irrigated by a sewage canal emanating from the Gurgaon city.

Sultanpur

Sultanpur is located 9 km from the Gurgaon city. The village is dominated by the Rajputs, numerically and in terms of land ownership. The village is not served by an irrigation canal and the groundwater is saline. Under these circumstances, farmers cultivate the less-water-consumptive crops, viz., pearl-millet, sorghum, wheat and mustard. This village is located adjacent to the Sultanpur National Park, a well-known wetland that was developed by acquiring land from this village. Sultanpur has seen land acquisitions for a variety of purposes over the last two decades; at the time of this research, further land acquisition was on the anvil for the development of an SEZ by Reliance Industries, a corporate giant.

Implications of Urbanization for Rural Water Use: The Findings of the Study

Peri-urban settlements tend to be at the receiving end of urban development and bear the brunt of the development of urban residential and industrial areas in many ways. In particular, pressure on water resources can come from many quarters; therefore, farmers' access to water for irrigation may be adversely affected as groundwater succumbs to other competing uses, such as industry, farmhouses, recreation and conservation. At the same time, access to water declines with the diminished availability of power that is diverted to meet the requirements of the city.

Competing Pressures on Groundwater Resources

The effects of pressure on groundwater resources from competing uses were particularly evident in Sultanpur. The extent of vulnerability as shaped by the limited availability of water described earlier has been aggravated by developments and changes in land use patterns around the village over the past decades. On the one hand, the mushrooming of farmhouses and the development of the Sultanpur National Park in 1971 has aggravated the stress on the village's groundwater resources; on the other hand, the erratic supply of electricity that is diverted in order to meet the growing requirements of the expanding city hampers the operation of tubewells.

There are about 60 farmhouses in the vicinity of the village, most of which extract water using submersible pumpsets. These farmhouses have been around since the mid-1980s. The water table depth is about 60 ft. A rough estimate of the annual pace of decline in the water table level, arrived at on the basis of discussions with the villagers, is 5–7 ft. As the water table has fallen, most of the small and marginal farmers are forced to leave their lands uncultivated since they are unable to dig deep enough. The erratic availability of power also interferes with pumping of water. This problem has become more acute in recent years as electricity supply has been diverted to meet the city's requirements. In the absence of reliable power supply, farmers are forced to pump water whenever power is available, rather than when their crops need it. The combined effect of both these factors is that many farmers are forced to leave their land fallow, or to cultivate only one crop in a year.

Location of Water Treatment Plants

The location of water treatment plants in peri-urban areas that supply water to the city also has had an adverse impact at the local level. The Basai water treatment plant that was established to supply water to Gurgaon city has been a mixed blessing for Basai's residents. On the one hand, it has made available drinking water to the residents of Basai and has provided irrigation to some; on the other hand, it has led to a rise in the local water table level, posing a threat to buildings in the region. Broken pipes and leaks from the water treatment plant have led to an increase in the mosquito population that causes several vector-borne diseases. With the rise in the water table level, the adjacent agricultural fields are known to be losing their productivity. Farmers are unable to grow wheat on this tract of land and it seemed that in the years to come, they would be unable to grow any crop at all. "HUDA has given water, but has not paid attention to other problems," was an opinion often expressed by farmers in field interviews.

Impact of Polluting Industries

Members of the rural communities often take advantage of the fact that the regulatory capacity of the government authorities is weak in peri-urban areas, particularly those that are outside the municipal boundaries (Parkinson and Tayler 2003). The location of factories near the *phirni* (boundary) of Shahpur Khurd village was identified as a perpetual source of noise and groundwater pollution; the untreated wastes from the factories found their way into the groundwater aquifers. These factories had been relocated from Delhi and were identified as a nuisance by peri-urban residents. Residents complained of a vibrating sensation in the ground throughout the day caused by their operation. They strongly felt that these factories should be located at least a certain distance away from the village, and particularly from religious places. Similarly, factories located around Basai village contaminate local village ponds by discharging their waste into them.

Links with Land: Demise of Local Water Management Institutions

The impact of the emergence of PUI was also felt on the CPR (Common Property Resource) institutions for water management, such as *johads* or village ponds. This was observed in all the four villages chosen for the study. This impact was felt most by the peri-urban poor and landless, as CPRs on which they depended for their livelihoods were diverted to urban use.

Basai village initially had six *johads*, of which three lay on the land acquired by HUDA for the development of residential sectors in Gurgaon and one lay on a piece of land acquired for the development of a public school.[3] An important group of people thus affected here were the potters who depended on *johads* for de-silting and on horses and livestock for carrying the clay to their places of work. With the takeover of both grazing lands and *johads*, the bases for their traditional livelihoods were eroded and they were forced to take up alternative occupations.

In Sultanpur village, among the Panchayat lands that were proposed to be acquired for the development of the Reliance SEZ was the land on which a water supply tank was built, managed and operated by PHED (Public Health and Engineering Department). This tank was the source of drinking water supply to much of the village. Besides, the acquisition of land for the construction of a highway inconvenienced peri-urban residents by affecting their routes and access to water sources. Since the local groundwater is saline, the residents of Sultanpur obtain water from a distance of about 1.5 km away by crossing a railway track. With the construction of the highway, they had to divert their route to the point of water collection and walk a longer distance.

Changing Locus of Control Over Village Resources

An important impact of the PUI is the shift in the locus of control over village resources to outside the village, as urban residents take part in the auction of village ponds and lands. Once again, this has implications for the livelihoods of those who depend on them.

In Shahpur Khurd, there were three *johads*. They primarily catered to the drinking water needs of the livestock. Earlier, these *johads* had been managed by the villagers through collective contributions of labor and capital. Subsequently, however, one of the three *johads* were auctioned off and the proceeds of the sale were retained by the Panchayat for use in rural welfare activities.

The *johad* was auctioned off to contractors predominantly from outside the village, particularly from Delhi, who have been using it for fishing. Though this auctioning of CPRs is an important source of funds for the village Panchayat, it also means that the locus of control over village resources has shifted outside the village.

A similar process was seen in Karnera. There were initially five *johads*, of which three were auctioned off to fisheries. After 1952, with the takeover of the *johads* by the Panchayat, the tradition of auctioning off the *johads* started. Before 1952, the villagers used to desilt it. Now, the Panchayat puts the *johads* on auction. The *pattedar*, or contractor, takes the *johads* on auction for a period of 2–3 years. The task of fishing is taken up on contract normally by Muslims. There are no Muslims in the village and these Muslims come from outside the village, again predominantly from Delhi. The *johads* are used for fisheries and for the cultivation of a fruit locally called *singhara* (water chestnut). The adjacent towns provide a market for fish.

An important impact of the practice of auctioning off the *johads* is that the *kumhaars* (potters) have lost their rights to a livelihood. Their traditional occupation has been pottery; however, they do not have access to clay any longer, which they used to get from the *johads*. The potters do not own agricultural land and now work predominantly as agricultural laborers.

Water Use in Peri-urban Agriculture

Peri-urban agriculture is most often not officially recognized as an urban land use, even though it is widely practiced in several areas (WII and IWMI 2006). Several characteristics of peri-urban agriculture in the villages studied may be noted. Perhaps, the most significant of these are the variety of irrigation sources, the prevalence of both formal and informal means of water allocation, the crucial role of peri-urban agriculture in supporting livelihoods and the use of wastewater.

Peri-urban agriculture has been understood to have an important role in providing employment to poor people in the fringe areas of Delhi (Marshall et al. 2003). Landless people are involved in peri-urban agriculture as agricultural laborers or as cultivators of leased land. Vegetable cultivation, in particular, is done mainly by farmers with low socio-economic status and having small or marginal landholdings. This type of vegetable cultivation supports livelihoods primarily through food provision, and income and employment generation.

Another notable characteristic of peri-urban agriculture is the diversity of sources of irrigation. In Basai village, for instance, depending upon the location of the agricultural fields, farmers can benefit from different sources of irrigation. In the absence of an irrigation canal, most farmers irrigate lands through private tubewells. There are two other sources as well. One is an underground pipe laid down to bring water from the HUDA water treatment plant to a temple in the village. Farmers whose fields lie along this pipe irrigate them by making a cut in the pipe and diverting water to their fields. Second is a sewage canal emanating from the Gurgaon city; farmers whose fields lie along this canal are able to irrigate using that source as well. From the perspective of irrigation, therefore, the best placed are those whose lands are geographically scattered so that they can irrigate them from different sources.

It is important to note, however, that while peri-urban residents may receive water for agriculture from a variety of sources, water supply for irrigation in north-western India is likely to be intermittent and uncertain. Partly, this is to do with the design characteristics of irrigation systems that are protective in nature, seeking to distribute a water supply thinly over a large area (Narain 2003; Mollinga 1998). Further, both statutory and non-statutory means of allocating water exist. There is a statutory *warabandi* schedule that defines a farmer's water rights and determines the pattern of resource allocation; however, farmers engage in an exchange of their time shares on the basis of mutual understanding and social relations (*bhaichara*). Water rights, though defined by the state laws, are, therefore, realized a normative system based on social relationships, pointing to the existence of legal pluralism (Narain 2003). This was noticed in Shahpur Khurd and Karnera, both of which are served by irrigation canals.

Another important source of irrigation for peri-urban agriculture is wastewater. Wastewater has a high potential for re-use in agriculture (WII and IWMI 2006). It offers an opportunity to increase food production and environmental security by avoiding direct pollution of rivers and surface water, conserving significant proportion of river basin water and disposing off municipal wastewater in a low-cost, sanitary manner. In the peri-urban context, sewage-irrigated agriculture enables farmers to overcome constraints to agriculture posed by poor-quality groundwater or absence of an irrigation canal, and thereby widen their cropping choices. The re-use of urban wastewater thus emerges as an important rural–urban linkage in the form of resource transfers from urban to rural areas, though it is known to have adverse health implications for consumers as well as growers of such produce.[4]

In the villages studied, sewage-based irrigation was found to be particularly important in Basai. The discharge of sewage from the Gurgaon city became an important source of irrigation for paddy cultivation that would otherwise not have been possible. However, the benefits of sewage-based agriculture were shared unequally among peri-urban residents, since access to wastewater depended predominantly on the location of their fields.

Conflicts and Social Mobilization

As competing pressures on water increase, rural–urban conflict has begun to manifest, particularly in Gurgaon.[5] In the third week of March 2008, farmers living near Gurgaon breached the Gurgaon canal that is the major supplier of water to the city, forcing the residents of Gurgaon to buy water from private sources (tankers) at prices as high as INR 5–700 per 5,000 litres.[6] About 400 water tankers had to be pressed into service to supply tubewell water to the people of Gurgaon on March 24, 2008; however, the water supplied by the tankers could meet just about 30 percent of the total demand for water. The water crisis in Gurgaon is seen as an outcome of the short-sightedness of the government in issuing licenses for the construction of malls and residential areas without taking cognizance of the water availability.[7]

There have recently emerged some efforts at mobilization around water. Most of these efforts are led by Civil Society Organizations (CSOs). For instance, in Gurgaon, Resident Welfare Associations (RWAs) have approached judicial authorities against the issue of licenses to builders and property dealers in the wake of the steadily diminishing water supply as manifest in the falling groundwater levels. Non-governmental organizations (NGOs) like Society for Urban Regeneration (SURGE) have been instrumental in constructing water harvesting pits in the district.

Issues for Governance and Public Policy: Overcoming the Rural–Urban Dichotomy in Development

As is evident in the previous discussion, urbanization processes are impacting the access of peri-urban residents to water in various ways. Generalizations are difficult to make, and a very localized approach is needed to examine the various ways in which urbanization affects peri-urban resident's access to water of a sufficient quality and quantity, as well as water management practices and institutions.

There has been much attention paid by the media lately to the issue of land acquisition and financial compensation to landowners in the process, in Gurgaon as well as elsewhere. However, the implications of these processes for access to water have received scant attention. In a situation where water rights are tied to land rights, the acquisition of lands for industrial and urban development implies *ipso facto* loss of access to water sources. When private agricultural lands are acquired, access to local groundwater resources is lost. Likewise, access to CPRs like village ponds is lost when the lands on which they are located are acquired.

Even as far as the acquired lands are concerned, the payment of compensation has thus far typically been to landowners while tenants and sharecroppers have not been part of a rehabilitation policy. Several studies have shown that it is the landless and poor who tend to depend much more on the CPRs including sources of water.[8] Thus, it is important that policies for urbanization and land acquisition include some measures for providing alternative sources of livelihoods to these landless and poor people or alternatives for the functions that the CPRs perform in a rural economy.

While a separation of water rights from land rights is imperative (there has been an ongoing debate in India on the subject), it remains difficult to operationalize. Besides, in a setting where contiguous tracts of land are acquired for such activities as the development of SEZs, this separation would have little significance unless peri-urban residents are also compensated for the water to which they lose access. Needless to say, assigning a value of compensation for the loss of access to water (i.e., groundwater beneath private agricultural lands or village ponds that are located on lands acquired for non-agricultural purposes) is methodologically, ideologically and operationally complex.

At a more fundamental level, challenges to peri-urban governance arise because of the fragmented approach to urban and rural development. Urban authorities define their mandate narrowly in terms of developing urban areas, not accounting for the consequences of such urban development on the rural areas. The mandate of HUDA, for instance, is defined in terms of the development of urban areas, even when it entails land acquisitions from rural areas. HUDA does not have, as part of its mandate, any mechanism to deal with the (rural) consequences of land acquisition for (urban) development, such as the loss of rural livelihoods or access to the CPRs.

Another important area for intervention, especially in the context of water use in peri-urban areas, is treatment of sewage water. A combination of measures to educate residents of peri-urban areas on and increase their awareness of the consequences of consumption of sewage-cultivated crops is needed along with improved infrastructure for the treatment of sewage to counter the health implications of sewage- based agriculture (IWMI 2006, 2003).

As India continues to urbanize, an important challenge will be to integrate planning for urban and rural development. The 74th Amendment to the Constitution of India (1992) provides for the creation of District Planning Committees (DPCs) and Municipal Planning Committees (MPCs). The mandate of these committees is the effective integration of rural and urban planning and spatial and economic development for the entire district (Brook et al. 2003). There is a need for creating such organizations on a larger scale to integrate and address the concerns of the PUI.

Improving synergies between local governments, NGOs, local civil society and the private sector can play an important role in supporting the positive aspects of rural–urban interactions while reducing their negative impacts (Tacoli 2002). The balance of evidence on the PUI world-wide points to the potential of local-level approaches. In peri-urban areas, which are in transition from rural to urban, have inadequate institutional cover and are difficult to bring directly within the purview of rural and urban jurisdictions. CSOs thus have an enormous potential to improve local environmental conditions, resolve political conflicts in governance and scale up environmental management activities (Dahiya 2003). There are several cases of local-level action in addressing the per-urban challenges world-wide; in India, they offer important lessons for scaling up and replicating.[9]

Conclusion

As India continues to urbanize and cities expand with an enlarging ecological footprint on the rural resource base, integrating rural and urban planning will be extremely important to mitigate the negative impacts of these. In the absence of a property rights structure for water or institutions for integrated water resource management across urban and rural areas, conflicts over water will continue to intensify. While some of these will be visible, others will be more tacit, such as the loss of sources of water as a result of land acquisition. Conventional approaches to urban and rural development will be ineffective in ameliorating the concerns of peri-urban residents in the absence of direct local-level action in which NGOs, CSOs and peri-urban residents themselves have a role to play. This requires the creation of platforms that bring together rural and urban governments, planning authorities and peri-urban residents.

As shown in this chapter, there is a variety of ways in which current patterns of urbanization impact peri-urban residents' access to water. Generalizations are difficult. The message of this essay is not that these changes are happening everywhere, but that similar processes are likely to unfold, as cities expand by engulfing lands from peripheral areas and impacting the access of peri-urban residents to water, as well as water management practices and institutions. Extremely localized approaches to analyze these changes are needed to build up on the process documentation, and participatory and ethnographic researches. These should then be used as a basis for devising local-level interventions mobilizing CSOs, NGOs, local governments and peri-urban residents.

≈

Notes

1. "Water Supply Project for Faridabad," *The Indian Express*, September 15, 2009, http://www.indianexpress.com/news/water-supply-project-for-faridabad/423740/ (accessed November 9, 2009).
2. "2nd Water Treatment Plant for Gurgaon," *The Times of India*, November 5, 2009.
3. The fifth one, located near the fields, was polluted with waste water on account of the discharges from one of the factories located in the vicinity. Subsequently, this was considered no longer fit for use by the livestock.
4. The role of wastewater irrigation in peri-urban agriculture across several cities of the world has been well documented. See, for instance, Bradford et al. (2003); Eaton and Hilhorst (2003); Parkinson and Tayler (2003); De Zeeuw and Lock (2000). See also IWMI (2006, 2003) and Feenstra et al. (2000).
5. Rural–urban conflicts on water are also witnessed in other Indian cities, like Chennai (Janakarajan 2009). Some efforts to resolve them have been made through the formation of multi-stakeholder platforms, sidelining the village Panchayats. Janakarajan notes that some threshold level of scarcity may be required in order to mobilize the peri-urban residents.
6. This was reported in "Private Tankers Take Gurgaon Hostage," *The Hindustan Times*, March 25, 2008, p. 1.
7. Meinzen-Dick (2000) notes that water has many values: social, ecological, religious, cultural and political. As the demand for water in cities has grown, cities look further and further afield for sources of water. This phenomenon of cities diverting water from other uses, notably agriculture, to themselves has been noted in cities like Kathmandu, Ahmedabad, Chennai and Los Angeles, as also in smaller towns and urban centers. Water transfers may be private, unplanned and ad hoc, with individual well-owners pumping water into tankers to be sold in the city, or public and planned, with water districts taking water from villages for selling it in the city, with or without compensation to villages. Meinzen-Dick further argues that we need to first understand the various claimants to water in

peri-urban areas, identify their uses and the various means adopted by them in order to meet their water needs; and the create processes for negotiation among the various stakeholders.

8. For a recent analysis of this issue, see Mishra et al. (2008).
9. See, for instance, Ahmed and Sohail (2003); Brook et al. (2003); Dahiya (2003); Dayaratne and Samarawickrama (2003).

References

"2nd Water Treatment Plant for Gurgaon," *The Times of India*, November 5, 2009.

"Private Tankers Take Gurgaon Hostage," *The Hindustan Times*, March 25, 2008, p. 1.

"Water Supply Project for Faridabad," *The Indian Express*, September 15, 2009, http://www.indianexpress.com/news/water-supply-project-for-faridabad/423740/ (accessed November 9, 2009).

Ahmed, N. and M. Sohail. 2003. "Alternate Water Supply Arrangements in Periurban Localities: *Awami* (People's) Tanks in Orangi Township, Karachi," *Environment & Urbanization*, 15(2): 33–42.

Bradford, A., R. Brook and C. Hunshal. 2003. "Wastewater Irrigation in Hubli-Dharwad, India: Implications for Health and Livelihoods," *Environment & Urbanization*, 15(2): 157–70.

Brook, R., S. Purushothoman and C. Hunshal (eds). 2003. *Changing Frontiers: The Periurban Interface, Hubli-Dharwad, India*. Bangalore: Books for Change.

Dahiya, B. 2003. "Hard Struggle and Soft Gains: Environmental Management, Civil Society and Governance in Pammal, South India," *Environment and Urbanization*, 15(1): 91–100.

Dayaratne, R. and R. Samarawickrama. 2003. "Empowering Communities in the Perurban Areas of Colombo," *Environment and Urbanization*, 15(1): 101–10.

De Zeeuw, H. and K. Lock. 2000. "Urban and periurban agriculture, Health and Environment." Discussion paper presented at the FAO-ETC/RUAF Electronic Conference on Urban and Periurban Agriculture on the Policy Agenda, August 21–September 30.

Eaton, D. and T. Hilhorst. 2003 "Opportunities for Managing Solid Waste Flows in the Periurban Interface of Bamako and Ouagadougou," *Environment and Urbanization*, 15(1): 53–63.

Feenstra, S., R. Hussain and W. van der Hoek. 2000. *Health Risks of Irrigation with Untreated Urban Wastewater in the Southern Punjab, Pakistan*. Lahore: Institute of Public Health, Lahore and International Water Management Institute (IWMI) Pakistan Program.

Government of India (GoI). 2007. *Groundwater Information Booklet: Faridabad District, Haryana.* New Delhi: Central Ground Water Board (CGWB), Ministry of Water Resources, GoI.

Halkatti, M., S. Purushothaman and R. Brook. 2003. "Participatory Action Planning in the Periurban Interface: The Twin City Experience: Hubli-Dharwad, India," *Environment and Urbanization*, 15(1): 149–58.

Iaquinta, D. L. and A. W. Drescher. 2000. "Defining Periurban: Understanding Rural-Urban Linkages and Their Connection to Institutional Contexts." Paper presented at the Tenth World Congress of the International Rural Sociology Association, August 1, Rio de Janeiro, Brazil.

International Water Management Institute (IWMI). 2006. "Recycling Realities: Managing Health Risks to Make Wastewater an Asset." Water Policy Briefing 17, IWMI and Global Water Partnership (GWP), February.

———. 2003. "Confronting the Realities of Wastewater Use in Agriculture." Water Policy Briefing 9.

Janakarajan, S. 2009. "Urbanization and Periurbanization: Aggressive Competition and Unresolved Conflicts: The Case of Chennai City in India," *South Asian Water Studies*, 1(1): 51–76.

Kundu, A. 2008. "Socio-economic Segmentation, Inequality in Micro-environment and Process of Degradation Peripheralization in New Delhi," in A. L. Singh and S. Fazl (eds), *Urban Environmental Management*, pp. 45–75. New Delhi: B. R. Publishing Corporation.

Marshall, F., R. Agarwal, Dolf te Lintelo, D. S. Bhupal, R. P. B. Singh, N. Mukhejee, C. Sen, N. Poole, M. Agarwal and S. D. Singh. 2003. *Heavy Metal Contamination of Vegetables in Delhi: Executive Summary of Technical Report*, http://toxicslink.org/docs/06102_Finding_of_Heavy_Metal_Contamination_of_Vegetables.pdf (accessed September 11, 2013).

Meinzen-Dick, R. 2000. "Values, Multiple Uses and Competing Demands for Water in Periurban Contexts," *Water Nepal*, 7(2): 9–12.

Mollinga, P. P. 1998. "On the Waterfront. Water Distribution, Technology and Agrarian Change in a South Indian Canal Irrigation System." PhD thesis, Wageningen Agricultural University, Wageningen, The Netherlands.

Mishra, A., N. Nayak, R. Ghate and P. Mukhopadhyay. 2008. *Common Property Water Resources: Dependence and Institutions in India's Villages.* New Delhi: The Energy and Resources Institute (TERI).

Narain, V. 2009. "Growing City, Shrinking Hinterland: Land Acquisition, Transition and Conflict in Peri-urban Gurgaon, India," *Environment and Urbanization*, 27(2): 501–12.

———. 2007. "A Tale of Two Villages: Transition and Conflict in Periurban Gurgaon." Management Development Institute (MDI) Working Paper Series 002, MDI, Gurgaon.

———. 2003. *Institutions, Technology and Water Control: Water Users' Associations and Irrigation Management Reform in Two Large-scale Systems in India.* Hyderabad: Orient Longman.

Narain, V. and S. Nischal. 2007. "The Periurban Interface in Shahpur Khurd and Karnera, India," *Environment and Urbanization*, 19(1): 261–73.

Parkinson, J. and K. Tayler. 2003. "Decentralized Wastewater Management in Periurban Areas in Low-income Countries," *Environment and Urbanization*, 15(1): 75–90.

Tacoli, C. 2002. "Changing Rural-Urban Interactions in Sub-Saharan Africa and Their Impact on Livelihoods: A Summary." Working Paper 7, International Institute for Environment and Development, London.

Winrock International India (WII) and IWMI. 2006. *National Workshop on Urban Wastewater: Livelihood, Health and Environmental Impacts in India: Proceedings.* New Delhi: United Services Institution.

Yin, R. K. 1984. *Case Study research: Design and Methods.* Applied Social Research Methods Series, vol. 5. Thousand Oaks: Sage.

12

Filtering Dirty Water and Finding Fresh One

Engaging with Tradition in Dug-Well Intervention in North Bihar

Luisa Cortesi

This chapter aims to critically analyze the concept of tradition[1] as often ideologically understood in the context of water management interventions in India. The argument engages with the following questions: What is tradition? Is it history? Is it internal and endogenous? Is it shared knowledge? Is it equitable? Is it sustainable? Is it opposed and disrupted by externally enforced intervention from the state? This article suggests that tradition is a category that needs to be unpacked, as a simplified notion of it might result in concealing its complexities and problems in terms of program coherence and feasibility.

This argument, however, does not take away the historical and current value of tradition; it discusses with the help of anthropological literature, that "tradition" is often used as a black box that carries unquestioned assumptions. The synthesis of a typical traditionalist discourse shows how the same is based on dichotomic assumptions and exemplifies some of the dangers that may arise from its use. Instead of proposing examples of water management interventions falling into what will be identified as common *traps* of traditionalist discourse, this chapter examines an example of a careful engagement with the concept. The example is provided by the case study of a dug-well revival intervention by Megh Pyne Abhiyan (MPA), a network of grassroots organizations working in water management in north Bihar. According to MPA's understanding,[2] tradition, as an incoherent and contested corpus and a delicate political arena for commercial and political interests, will open up a way of relying on ethnographic research and on a specific modality of considering the issue of knowledge and the role of experts. The argument is that, in order to build constructive occasions of dialog, it is relevant to reflect on actual practices of development that take on board anthropological knowledge.

This chapter is highly indebted to Foucaulian notions (Foucault 1972, 1977, 1980) of discourse, power, representation and truth. Three concepts by Foucault, discourse as a system of representation; knowledge linked to power that not only assumes the power of the truth but makes itself the truth; and discursive formations (for example, history, nature and tradition) that sustain a regime of truth to give power a non-contestable form – are the theoretical background for this chapter. However, the aim of this work is to practically contribute to the debate on tradition in the sector of water management. With this concern in mind, the theoretical approach will not be made explicit throughout.

Traditionalist Discourse

The starting point of this chapter is to highlight a (non-exhaustive) series of conceptual traps in which water management interventions may fall into while taking on board the idea of tradition in project design and implementation. The "traditionalist discourse" of natural resource management will be briefly outlined and contradicted with reference to the anthropological literature. As a response, the case study of an alternative approach of engaging with "tradition," both on conceptual and practical levels, will be illustrated.

The main argument of this section is that "tradition" is considered an unproblematic category by a "traditionalist discourse" that adopts a chain of assumptions as connected with the concept. The "traditionalist discourse," as referred to by Sinha et al. (in Mosse 2003),[3] used in water management can be synthesized in the following manner. In the past communities, developed traditional systems to manage water (or natural resources in general) were in harmony with nature. An example can be interventions regarding tank systems, considered a sustainable and equitable way of managing water since ancient times in Tamil Nadu and other states in India, whose rejuvenation is overdue in order to strengthen livelihoods (e.g., Annamalai 2006; Arati 2006). I should state here that my point is not to criticize programs aiming at rejuvenating tank irrigation systems, but simply to provide an example of an ideology of development that we can refer to as "traditionalist discourse," of which tank rejuvenation programs in India offer a suitable example.

The traditionalist discourse, which I have defined as an unproblematic chain of assumptions logically connected with the concept of tradition in natural resource management, presents several postulations that are interesting to examine from a sociological point of view. Here, I will identify what I consider the most heuristically relevant ones in order to discuss them in the light of anthropological literature in the next section. First, tradition is considered a part of the history of a community, as the community established or developed it in an undermined past. Often, the history of tradition sees its disruption by the state, more specifically by the colonial state in the Indian context. Second, the community has developed a particular tradition/system autonomously and endogenously. Third, the community is supposed to collectively uphold the tradition-related knowledge and guarantee its unadulterated transmission to the future. Fourth, community, conceived of as a homogenous group governed by just and impartial relations and rules, ensures equitable access to natural resources. Fifth, the concept of "harmony with nature," also phrased otherwise, can be commonly translated into contemporary discursive terms as a sustainable management of natural resources. Moreover, the highlighted assumptions of the traditionalist discourse – tradition is endogenous and authentic; based on collectively shared knowledge; unbiased and, therefore, ensuring equitable use of natural resources by community members; in harmony with nature and, therefore, guaranteeing sustainability; and easily restorable – can be associated with a parallel and opposite assumption of exogenous interventions, decided forcibly by the state, disrupting the harmonious relations between the community and the natural environment, and hence not manageable, equitable or sustainable. This dichotomy will be further explained later in this chapter.

Anthropologic Literature and Tradition

The anthropological literature helps to unravel the assumptions behind the traditionalist discourse and identify the related traps in which natural resources management programs may fall into. Let us examine the assumptions highlighted earlier in the light of anthropological literature.

First, several ethnographic studies reveal that historical authenticity should be demonstrated with historical research. Going back to the example of tank irrigation, Mosse's contribution, for example, denies the typical historical timeline of tank system, supposedly flourishing in the past and disrupted by the colonial state. Mosse (1999, 2003) analyzes the history of rural tanks in Sivaganga district in Tamil Nadu and concludes that there is no decisive moment and reason for tank irrigation to collapse. He states, "Tank systems have, in fact, been interpreted as being in a state of decline, neglect and disrepair wherever they have been described" (Mosse, 1999: 307). Moreover, not all the tanks collapsed when the colonial rule started, so their history must have differed from case to case. Anthropology calls for historical evidence to substantiate what I highlighted as the first assumption of the traditionalist discourse, the authenticity of tradition.

Second, what is reported as endogenous may have become as such through people's narratives. Aubriot's study (2004) of Aslewacaur village in Nepal points out that the ancient irrigation system, claimed to have been established by "the ancestors," was probably imported from north India. Similarly, Shah (2003) remarks that endogenous and exogenous are not so opposed and their supposed distinction is often indistinct. In all the "endogenous" tanks she examines in her ethnographic study, only hybrid seeds of paddy, not endogenous varieties, are grown with the help of chemical fertilizers and pesticides. Moreover, the history of cultures and artefacts is full of examples of adoptions, acquisitions, transformations of material uses, symbolic values and functioning within the relationships between people and the concerned natural resources. Sagant's agricultural history of Nepal talks about the importation by immigrants of ploughing and rice transplantation, together with cattle and goat breeding (Aubriot 2004). Agricultural life was changed, overturned and maybe also uniformed by their arrival, but in Aslewacaur, Aubriot admired, in the time frame of her fieldwork, a sophisticatedly organized and efficient agricultural system.

Third, the extensive anthropological debate on the concept of knowledge shows how knowledge is contested, not collectively held but differentiated in multiple voices (Sillitoe and Bicker 2003). Moreover, as Novellino (2003) adopting a Foucaultian ap-proach argues, local knowledge, and hence tradition, is neither easily accessible by anthropologists (and development practitioners) nor necessarily translatable to their language. In addition, in the situation of a developmental encounter, the transmission of knowledge can be manipulated for given interests. In fact, the copious debate on community participation in natural resources management interventions examines and ethnographically exemplifies not only how development interventions have failed to grasp or politically contextualize local knowledge (White 1996), but also questions the way in which the interventions themselves are being strategically appropriated (Green 2000; Pigg 1992). From a theoretical point of view, Hobsbawm's work on the invention of tradition (Hobsbawm and Ranger 1983), together with Appadurai's notion of the past as a plastic symbolic resource (Mosse 2003), suggests that tradition can be manipulated as a symbolic capital for political convenience.

Fourth, this understanding of community recalls the Weberian contraposition between *Gemeinschaft* and *Gesellschaft*, and allows an idealized and dichotomic use of the concept. Instead, the literature highlights state-made communities (Mosse 1999), communities as ideologies, and strategic communities (Pigg 1992; Li 1996) For example, the traditionalist view of the abandonment of tank irrigation is based on the idea of the fundamental opposition state-versus-community, and at the same time on the association community-tank-tradition. Gupta (1998) demonstrates the inaccuracy of the antinomy state–community, using the concept of "blurred boundaries." In his view, the state is part of the society, the society is made by the state, and people are active members of both. Following Mosse (2003), tanks, instead of being symbols of autarchy and tradition, are often the infrastructures that make the state locally visible. More generally speaking, what is portrayed as traditional irrigation managed by the community has been often a powerful idea of the state to govern ensuring law and order (ibid.). Aubriot asserts: "Even when farmers present their rural system as indigenous, and set by their ancestors, the relation with the state cannot be excluded" (2004: 17). More importantly, community and more specifically Indian communities are not specifically homogenous, nor have they been ensuring equitable access to resources. In the era of development, community-based programs often distribute the benefits of the program over unquestioned inequalities, hence reinforcing pre-existing economic disparities (Guggenheim 2003).

Fifth, the concept of sustainability would require a specific and lengthy discussion The anthropological literature calls for the need to question "nature" and the status of "being in harmony with nature," accusing these ideas for being part of an ideology of the past (Williams 1973). Sinha et al., more specifically, counter the simplistic claimsthat "women, forest dwellers and peasants were primarily the keepers of a special conservationist ethic," while "colonialism, modernity and development were exclusively responsible for the degradation of nature in India" (1997: 65).

The anthropologic literature further highlights that, not only is this traditionalist discourse historically incorrect, but behind these idealized and ideologized constructions of tradition, state and community, there are strategic agendas (Mosse 2003). Mosse argues that, while in the past *kudimaramath* (the tradition of tank irrigation repairs work by villagers in Tamil Nadu) was an "ideological argument enabling the assertion of state control (over water), while delimiting state obligations" (1999: 313), in the present, development agencies idealizing the pre-colonial past and custom, "selectively endorse particular social theories in constructing a rural society which is manageable in terms of present policy goals and administrative constraints" (ibid.: 306). While accepting that a simplistic notion of tradition could be motivated by political agendas and can result in counter-productive policies, I attempt to present the agent and the ethnographic effort that has enabled the case of a different reading of the concept of tradition.

MPA and Drinking Water in North Bihar

After presenting the anthropological view on the loopholes of the traditionalist discourse, I will examine more closely an experience of practical intervention of water management,[4] by a network of NGOs working in North Bihar. Therefore, the aim of this section is to introduce a subject, MPA, as the agency designing and implementing the intervention I will discuss, and to briefly sketch the contest over drinking water in north Bihar.

MPA in 2007–08 defines itself as a "belief in and commitment to inspiring a behavioral change amongst the flood-affected rural population, in order to foster their resilience and improve their individual and collective well-being". The network's motivation is to "construct a congenial social environment by stimulating cooperative action and accountability towards shared goods and problems through sustainable technological innovations and adaptations of conventional wisdom". MPA's endeavor is to "facilitate adoption of local strategies to ensure access to safe drinking water and improve the socio-economic conditions and health of the flood-affected population". MPA's commitments are put in practice both in the form of a campaign involving people in the issues of water and livelihood, and as a "functional network of grassroots organizations," viz., Gramyasheel, Kosi Seva Sadan, Samta, Ghogardiha Prakhand Swarajya Vikas Sangh and Savera, working in five districts across the northern part of the state, Supaul, Saharsa, Khagaria, Madhubani and West Champaran.[5]

It is relevant here to mention that as an anthropologist specializing in water management, I have been involved with MPA till the end of 2008, wherein I have shared the responsibility to define its conceptual basis and design its programmatic interventions.The purpose of this chapter is precisely to discuss how MPA, drawing on its expertise in water management and on its anthropological awareness, conceptually designed its programmatic interventions by engaging with the concept of tradition. In fact, this discussion aims to contribute toward enhancing the dialog between academic knowledge and practical commitment.

The northern part of the state of Bihar in east India, as divided by the Ganga river, is the playground of eight major rivers originating from the Himalayas. In north Bihar, flood is a recurrent disaster that, apart from taking away thousands of human lives almost every year, together with livestock and assets worth millions, has multiple, wide-ranging, direct and indirect consequences for various aspects of people's lives. In this context, the brutal problem of inaccessibility of safe drinking water arises during floods and translates into a myriad of health-related problems. Moreover, during the floods, groundwater accessed through hand-pumps is affected by different types of contaminants – biological and chemical (ammonia) ones from human excreta and a high (i.e., substantially above the permissible level) content of iron and arsenic – sporadically coupled with hardness and high pH value. In addition, water-logging in large tracts, as well as the lack of basic hygiene and sanitation facilities, worsen the health of people living in rural areas, as also adversely affect their socio-economic conditions. However, the long exploratory qualitative survey conducted in 2006[6] and the 2007–08 ethnographic fieldwork highlight and substantiate a general carelessness towards water quality by local inhabitants, growing disinterest in the practices of ensuring clean water and managing its availability at the community level, and high dependence on relief operations and its material benefits.

Admittedly, this brief and rather simplistic description of water problems in north Bihar is severely limited by the impossibility of providing an account of the related socio-economic situation and its political facets. Nonetheless, I need to introduce the intervention of MPA by presenting the most popular and widespread technology to access water in the area, i.e., the hand-pump, through a concise sketch of ethnographic findings.

Approximately, since the 1970s, drinking water in rural areas of north Bihar has been primarily accessed through hand-pumps. The technology, when introduced, faced a stiff resistance driven by symbolic and religious reasons, but the strategy of propagation it was such that soon the technology started gaining popularity at the household and village levels. At the same time, a modernist discourse, hinging on "modern" and "scientific"concepts of water quality, targeted the technology of dug-well as "backward" and the water accessed through hand-pumps as "safe" and "modern."[7] This phenomenon was, in turn, exploited by local politicians who used the hand-pump as a token to strengthen their relationship with their constituencies, by market forces for commercial interests, and by the development sector in order to perpetuate their patronage and display their achievement in providing safe drinking water sources to communities.

Moreover, depth is the criteria popularized by the bore drillers and government officials as the reliable parameter for accessing safe water, clearly indicating the intent of increasing sales and meeting targets without any concern for the hydro-geological dynamics and its impact on the quality of groundwater. In fact, hand-pumps have been suspected of being one of the possible reasons for the release of arsenic in aquifers because their drilling breaks young alluvial rock deposit strata and allows the passage of oxygen to permit the chemical reaction at the origin of the dissolution.[8] As a by-product of the politics of hand-pumps, instead, the household access to this facility has become a prominent symbol of social status. More precisely, the depth of the hand-pump is taken in consideration by the bride's family during marriage negotiations as a measure of the standard of life of the receiving family.

Finally, as reported earlier, the quality of the water accessed through hand-pumps is progressively becoming a contested issue, being confronted, although still rarely, by women on the basis of parameters like taste, smell and physical appearance. Men, parti-cularly family heads, instead defining the hand-pump water through the supposed depth of drilling. As a consequence, the overall social impact of hand-pumps is being understood in this research paper as an effect of their role as agents of transformation in the relations between people and water. In fact, the technology fosters an individual approach to water for three main factors: first, no requirement for collective management; second, need for investment in cash and reliance on external market and knowledge both for drilling and for repairs; and third, loosening of occasions for dialog and deliberations on and around water.

Although interconnectedness, as mentioned earlier, is not free from inequities, the observed consequence goes toward a fragmentation of collective mechanisms for the management of common goods and concerns.[9]

Given its mandate and mission, MPA has reacted to these findings by proposing to extend the research to the history of water management in the region. As a result, the research focus has moved towards the technology of dug-well. In the following months, the research findings instigated the design of a related intervention, as explained in the following section.

Dug-well Revival

Given the relevance of the concept of tradition in the field of natural resource management, put forward by several institutions and environmentalists since the 1980s, for example A. Agarwal and S. Narain (1988), it is impossible to escape from engaging with the concept of tradition and with the baggage of assumptions aggregated later on through the traditionalist discourse. The purpose of this chapter, as it will be clear in this very paragraph, is not to dismiss tradition and thereby throw the baby out with the dirty water, but instead to filter the concept of tradition from the traditionalist discourse and allow it in fresh and productive utilizations. My argument is that the way out of the loopholes of the traditionalist discourse precisely lies in taking the concept on board appropriately and engaging with it at the level of program design and implementation. With this objective, I portray the case study of a programmatic intervention of MPA, viz., dug-well rehabilitation.

At the outset, let me specify that any discussion of an intervention in the field of social development, including natural resource management, involves debates on several issues that cannot be dealt with here for lack of space. In fact, while discussing the conceptual structure of the program, several "buzzwords" and related sociological topics that deserve to be examined, will be instead simply mentioned. Also, this article is not concerned with explaining the scientific reasons behind the propagation of dug-wells as sources of clean drinking water when managed appropriately. Instead, the interest is in elucidating the way in which the designing of the intervention engages with the concept of tradition. For heuristic reasons, the argument will be organized to reproduce the structure employed in the preceding paragraphs.

First, as far as the historical relevance of the dug-well is concerned, specific ethnographic data collection took place for several months prior to the intervention's design.[10] Special attention has been paid to differentiating the actual perception of dug-well history by rural people across generational lines. While the last generation alive has expressed the relevance of dug-wells as the main source of safe water when appropriately cleaned, it has also recognized the shared management of this common property resource as difficult. Moreover, the generation of people born after the installation of hand-pumps, while not apparently interested in and aware of the dug-well technology, demonstrated to have inherited an attitude of valuing the dug-well as an asset and as a status symbol. In fact, even abandoned dug-wells taken into consideration during partition of landed property are prone to become the bone of contention in litigations. Moreover, considerable value is attached to dug-wells as a tradition from a religious perspective, as dugwell water is used for rituals in several temples. On the other hand, it is known that the dugwell is the only source of water which can help in the catastrophical event of fire burning down the village, since the hand-pump cannot source a large amount of water in limited time.

Second, focusing on the actual meaning and relevance of what can be considered tradition, MPA has given importance to the actual composition of internal and external forces, symbols and interests. Instead, it is not interested in constructing a precise historical account of dug-well use, nor in establishing the origin and source of the tradition. The outcome of the data collection procedure led to the recognition of dug-well as a locally meaningful symbolic repertoire of know-how, which, however, became contested and relegated as backward with the arrival and establishment of the hand-pump. The politics of water, involving water as a token gesture with the objective of garnering votes during elections, together with commercial interests, necessitated overcoming the supposed backwardness of the local dug-well. The other contributory factor behind the portrayal of dug-well as a backward technology was the growing relevance of the hand-pump as an economic status symbol in forging marital alliances, and as political status symbol in its being the recipient of patronage and relief benefits. It is in this sense that the inevitable composition of external and internal, symbolic and material forces centred around water have been considered noteworthy in our ethnographic study.

Consequently, the dug-well revival intervention by MPA was designed to make sure that small groups of neighbors are collectively and autonomously responsible for the management of the resource, hence fostering the concept of self-reliance in terms of clean water access. At the same time, MPA has not opposed or diminished the role of the state and has instead organized meetings to include Panchayat,[11] block and district representatives, together with other social forces, such as other local organizations and media representatives. On the other hand, MPA has refrained from allowing itself to be appropriated by vested political interests. In practice, dug-wells can be managed by the people themselves, without relying on external forces and without giving space to political manipulations. In fact, in case of a similar large-scale project being implemented, local conflicts may rise precisely because the low-cost solution diminishes the need for political support, hence overcome the standard procedure of patronage and power relations. For what the project implementation is concerned, the well-established organizations, run by committed local people, have acted as counterforce in the process of strategic appropriation of the program's resources. The problem of transmission of knowledge has also been considered by the project design. This issue will be specifically addressed in the following paragraphs.[12]

Fourth, equity is an important concern of MPA. The communities settled in the area, although organized by caste, are far from homogenous. MPA, assuming that involving community per se could hide power discrepancies, understands that explicitly addressing equity and caste stratification would not bring about a solution to the problem of social inequity in access to water resources. As a consequence, the program has been designed to be a collective event of re-appropriation of the water resource, in which a generic neighborhood is invited to intervene. The dug-well is cleaned and rehabilitated by the group of people living around it, at their expense. MPA participates in organizing the event, testing water scientifically and partially contributing, in case major repair work needs to be undertaken. The constant participation of MPA, its implicit Gandhian identity and its continuous presence through its grassroots organizations, together with the event of dug-well revival and the slogan written on the platform, are mainly measures to ensure collective use of dug-wells and prevent individual ownerships or disparity of access to water. In addition, the maintenance of the technology itself requires a repeated collective effort; hence, it concretely reassures the collective use of dug-wells, more then what formal rules can enforce. Moreover, the "campaign" is a recognized modality of social action resulting in enjoining of the duty to spread the message to by-passers and, therefore, allow access to dug-well water to outsiders.

Finally, dug-well is a demonstrable sustainable technology per se, which does not contaminate groundwater but naturally filters it from eventual pollutants present in the aquifer itself, and allows the use of shallow water at the pace of natural recharge, being connected but also independent of the systems of flood water control found in the area under study. The revival of dug-wells does not interfere with other systems of managing water, nor necessarily involves other practices, symbols or materials that require any further specifications. However, MPA ensures long-term commitment, both as a network and through its grassroots organizations, to maintaining a constant check on the social and environmental conditions of dug-well functioning and water management.

To sum up, MPA has promoted dug-well revival pilot projects not because dug-well is a "traditional" technology, but because it is driven by its mandate of sustainable and equitable water management, inspired by the related cultural symbols, material functioning and the local socio-environmental history. Dug-wells have been valued as locally significant, respectful of the environment and compatible with equitable socio-political progressive tendencies. In fact, the modality of their revival and management, as has been partially illustrated, has been designed in order to interpret the progressive changes inspiring the project.

Knowledge and Experts

After illustrating the intervention in the light of the discussion on traditionalist discourse, I would like to address a more focused question about knowledge and the consequent role of experts that allows us to bring forth the discussion on tradition as interpreted by MPA. Roth (2006), while maintaining a distinction between endogenous and external irrigation systems in his case study in Indonesia, points out the problem of external knowledge. In his words, "in the former [endogenous], technical, normative and organisational design principles [are] originated from system users and their socio-cultural environment. In the latter [exogenous], users are not designers. Irrigation design and development are based on external inputs and interventions" (ibid.: 32). Despite having already addressed the false dichotomy between external and internal, exogenous and endogenous, I can very well acknowledge the plea for a system that is decided by or through its users. More specifically, recognizing technology as a socio-technological artefact with politics (Pfaffenberger 1992; Shah 2003;), attention should be paid to the dangers of its design (Winner 1993), the politics inserted in the design itself, and the consequences for its users.

This article cautions against the use, in developmental interventions, of tradition as a black box, a meaningless paradigm, at times hiding covert interests and opening up conceptual and practical traps. At the same time, it argues for the relevance of the concept of tradition in terms of a reaction against imposed external and unproblematic technology-transfer-based interventions, as per the debate introduced by Anil Agrawal in the 1980s (Agrawal and Narain 1988). MPA's conceptual basis, indebted to his work, has employed the instrument of ethnographic research in order to gain an in-depth understanding of tradition in the context of water management interventions in north Bihar.

As explained earlier, tradition should not be assumed as a given and a holy corpus of coherent practices, but as an arena of political and sociological action. Therefore, MPA decided targeting its action at stimulating discussion on equitable interventions, both at the field level and, more broadly, in the sector of development.

At the field level, specifically, the intervention of dug-well revival has not been designed as a prescribed package of deliverable solutions, nor has MPA presented itself as an expert or decision-maker on water management. Instead, its role has been to provoke and facilitate an informed dialogue at the local level on water issues. Several practices have been devised with this goal. First, MPA has established an identity against relief and against delivering benefits. In fact, the identity-maker activity of the campaign, rainwater harvesting, has been conducted since 2007 without providing goods or material benefits to the people. The polythene sheet used for demonstrating rainwater harvesting is withdrawn from the villages with almost no economic resources employed. Second, all programs have been devoting specific attention to the "software," so to speak, to social dynamics instead of technological artefacts. This has been pursued through constant and multiple interactions and activities in order to initiate further discussion on the concerned issue. Third, all interventions focus on encouraging local innovation and transformation of the activities with the idea that constant renegotiation of knowledge is necessary in, and for, adaptive context. For instance, rainwater harvesting has been adopted in several different ways by groups and individuals, far beyond the imagination of MPA's practitioners. Fourth, very local social, economic, cultural, political, and environmental characteristics are well known to the practitioners involved in MPA, so that contextual interactions can happen smoothly. While these practices of interventions were shared by all the organizations belonging to MPA, very careful mentoring was required in order to keep the everyday operations so distinct from the typical development interventions' activities. Interestingly, it seemed that the most favorable reaction to this innovative modality was from the lowest socio-economic group, generally the last recipient in the distribution chain of relief activities.

More specifically, the dug-well revival activity has been based on the following three stakes. First, the intervention is designed in order to be an occasion for discussing knowledge regarding water quality, floods, dug-wells, hand-pumps, equity, long-term access to water. Second, the organization employs local people from nearby areas, trained to conduct water testing and to understand their working area following a non-intrusive modality based on the concepts of equity and sustainability. Third, no recipe for cleaning of dug-wells is provided, only facilitation of inter-generational transmission of local knowledge, considered as disrupted by the advent of hand-pumps and the intervention of market forces and political interests.

The modality of intervention illustrated till now is to be further explained. As ethnographic research becomes the basis for the project design, this modality contests in practice the project cycle typical of developmental interventions. On the one hand, for what the dug-well revival is concerned, the level at which the research informed the project and disrupted the standard project cycle is testified by the fact that no budget was allocated for the intervention, and the pilot project of 13 dug-wells took place relying upon the grassroots organizations' savings. Similarly, the team continued to engage in data collection and critical evaluation during the implementation stage, and the research output continued to inform different phases of the project evolution. It is relevant to specify that MPA decided to take an anthropologist on-board precisely for this reason, and is one of the few but increasing number of organizations valuing this kind of expertise, instead of relegating it to pre-implementation feasibility studies and post-implementation quick evaluations (Pottier 1993).

Conclusion: Anthropological Knowledge and Development

As it has been illustrated earlier, MPA's understanding of tradition is not one of a sacrosanct and coherent practice. Instead, it considers tradition to be a contested and constantly renegotiated corpus of incoherent and multiple knowledges. Moreover, MPA's engagement is with respecting the local history and the social ecology of the area of intervention, together with adhering to the values of the organization, i.e., interpreting progressive forces of change for equitable, sustainable and contextual water management practices.

In other words, MPA gives importance to applied sociological research on traditional water management practices not for the sake of historical analysis or in order to argue for the originality of those systems. Its aim, rather, is to enable the recognition of the actual relevance of these systems to the local context, their feasibility in the current economical and environmental contexts, and their value in the web of socio-political relationships. Therefore, ethnographic research becomes an instrument to keep a project "safe" from the traps of the traditionalist discourses, both in terms of understanding a tradition and in proposing a related intervention.

However, in conclusion, it is important to acknowledge two limitations of this approach. The kind of engagement I have been presenting is the result of a rare organizational effort, and at the current stage of the development sector, not sustainable in the more realistic sense of the term. Although it is not expensive economically, this inclination for the process, more than for supposedly quantifiable results, is not appreciated by funding agencies. Bluntly, this complexity scares funds. Cynically, we have to admit that this can be translated as financially not "sustainable". As a consequence, despite the refined practices and thinking behind MPA operations, the lack of funds is likely to push the organization to accept compromises and standardize its interventions.

Second, anthropological knowledge is difficult to be brought on board by development practice. While several academicians have addressed the questions of the social sciences' place in development (e.g., Grillo 1997; Gardener and Lewis 1996; Hobart 1993; Pottier 1993, etc.), further work is needed in order to bridge theoretical understanding and academic agendas with the policy and practice of development. However, the development sector (at least for what water management in India is concerned) is extremely resistant to accept social scientists as either researchers or practitioners. As an anthropologist in two different shoes, of academia and development, and in the light of the previous discussion, I am convinced about the possibility of a productive relation between academia and development, and would advocate for incentivizing mechanisms of reciprocal benefits. Social research should support development interventions as well as evaluation practices. Hence, this paper argues for ethnographic efforts that combine the agendas and constrains of academia and scholars on the one hand and those of development sector and practitioners on the other, in the quest for donors and academia to converge their resources and meet their social roles.

≈

Notes

1. Tradition is used in its conceptual form throughout, even if it is not accompanied by inverted commas.
2. The reference to MPA's approach refers to the time in which this article has been written, November 2008.
3. Mosse (2003) to whom I am highly indebted in writing this chapter, also cites Greenough's "standard environmental narrative."
4. While this argument has been forged in terms of water management, it is to be noted that no difference is acknowledged in terms of the purposes of the project, for example, irrigation or drinking water.
5. These self-defining sentences were used on leaflets, business cards and other communication material distributed in 2007 and 2008.
6. Fieldwork was conducted from May 2007 to September 2008.
7. It would be interesting to develop a parallel with Pigg's (1996) findings in Nepal.
8. In case of As2S3 (Arsenic Trisulfide) and in certain conditions. In fact, the mechanisms of oxidation and reduction of arsenic compounds and their mobilization in water are being studied and considered contradictory issues. See, for example, Harvey et al. 2005 and Piyush et al. 2002.
9. Similar findings are shared by Appadurai (1990).
10. Although it has not been feasible to proceed with archival study for lack of time and resources, government sources have been consulted, and primary importance has been given to ethnographic fieldwork in the extended area under study, which included several procedures of data verification and triangulation. Apart from the careful methodology, in order to increase the reliability of data collection, the ethnographic study was conducted across five districts of North Bihar.
11. Village council or the smallest unit of local governance.
12. The description of the project refers to its first phase in 2008.

References

Agrawal A. and S. Narain. 1997. *Dying Wisdom: Rise, Fall and Potential of India's Traditional Water Harvesting Systems.* New Delhi: CSE.

——. 1988. *Towards Green Villages: A Strategy for Environmentally-Sound and Participatory Rural Development.* New Delhi: Centre for Science and Environment (CSE).

Annamalai, V. 2006. "Rejuvenating Tanks for Sustainable Livelihood: Innovative People's Initiatives in Tamil Nadu." Paper presented at Tank Rejuvenation Workshop, 3–4 August, Institute for Social and Economic Change, Hyderabad.

Appadurai, A. 1990. "Technology and the Reproduction of Values in Rural Western India," in F.A. Marglin and S.A. Marglin (eds), *Dominating Knowledge: Development, Culture, and Resistance*, pp. 185–216. Oxford: Clarendon Press.

Arati, D. 2006. "Investigating the Impact of Cultural Action in Sustainable Tank Rejuvenation: Case Study of Svaraj/Oxfam India's Model in Nagarakere, Arkavathi Sub-catchment." Paper presented at Tank Rejuvenation Workshop, 3–4 August, Institute for Social and Economic Change, Hyderabad.

Aubriot, O. 2004. *L'Eau, Miroir d'une Société: Irrigation Paysanne au Nepal Central. Monde Indien, Science sociales.* Paris: CNRS Éditions.

Foucault, M. 1980. "Two Lectures," in C. Gordon (ed.), *Power/Knowledge: Selected Interviews*, pp. 78–108. New York: Pantheon Books.

——. 1977. *Discipline and Punish.* New York: Pantheon Books.

——. 1972. *Archaeology of Knowledge.* New York: Pantheon Books.

Gardener, K. and D. Lewis 1996. *Anthropology, Development and the Post-modern Challenge.* London: Pluto.

Green, M. 2000. "Participatory Development and the Appropriation of Agency in Southern Tanzania," *Critique of Anthropology*, 20(1): 67–89.

Grillo, R. 1997. "Discourses of Development: The View from Anthropology," in R. Grillo and R. L. Stirrat, *Discourses of Development: Anthropological Perspectives*, pp. 1–33. Oxford: Berg.

Guggenheim, S. 2003. "Crises and Contradictions: Understanding the Origins of a Community Development Project in Indonesia." http://www.cultureandpublicaction.org/bijupdf/guggenheim.pdf (accessed January 20, 2005).

Gupta, A. 1998. *Postcolonial Developments: Agriculture in the Making of Modern India.* Durham and London: Duke University Press (Chapter 1).

Harvey, Charles F., Christopher H. Swartz, Abu Bohran M. Badruzzaman, Nicole Keon-Blute, Winston Yu, M. Ashraf Ali, Jenny Jay, Roger Beckie, Volker Niedan and Daniel Brabander. 2005. "Groundwater Arsenic Contamination on the Ganges Delta: Biogeochemistry, Hydrology, Human Perturbations, and Human Suffering on a Large Scale," *Comptes Rendus Geosciences*, 337(1–2): 285–96.

Hobart, M. 1993. "Introduction: The Growth of Ignorance?," in M. Hobart (ed.), *An Anthropological Critique of Development: The Growth of Ignorance*, pp. 1–23. London: Routledge.

Hobsbawm, E. and T. Ranger (eds). 1983. *The Invention of Tradition.* Cambridge: Cambridge University Press.

Li, T. M. 1996. "Images of Community: Discourse and Strategy in Property Relations," *Development and Change*, 27: 501–27.

Mollinga, P. 2003. *On the Waterfront: Water Distribution, Technology and Agrarian Change in a South Indian Canal Irrigation System.* Wageningen University Water Resources Series. New Delhi: Orient Longman.

Mosse, D. 2003. *The Rule of Water: Statecraft, Ecology and Collective Action in South India.* New Delhi: Oxford University Press.

——. 1999. "Colonial and Contemporary Ideologies of 'Community Management': The Case of Tank Irrigation Development in South India," *Modern Asian Studies*, 33(2): 303–38.

Novellino, D. 2003. "From Seduction to Miscommunication: The Confession and Presentation of Local Knowledge in Participatory Development," in J. Pottier and P. Sillitoes (eds), *Development and Local Knowledge: New Approaches to Issues in Natural Resources Management.* London: Routledge.

Pfaffenberger, B. 1992. "Social Anthropology of Technology," *Annual Review of Anthropology*, 21: 491–516.

Pigg, S.-L. 1996. "The Credible and the Credulous: The Question of 'Villagers' Beliefs' in Nepal," *Cultural Anthropology*, 99(2): 160–201.

——. 1992. "Inventing Social Categories through Place: Social Representations and Development in Nepal," *Comparative Studies in Society and History*, 34: 491–513.

Piyush K. P., S. Yadav, S. Nair and A. Bhui. 2002. "Arsenic Contamination of the Environment: A New Perspective from Central-East India," *Environment International*, 28(4): 235–45.

Pottier, J. 1993. *Practicing Development: Social Science Perspectives.* London: Routledge.

Roth, D. 2006. "Which Order? Whose Order? Balinese Irrigation Management in Sulawesi, Indonesia," *Oxford Development Studies*, 34(1): 33–46.

Shah, E. 2003. *Social Design: Tank Irrigation Technology and Agrarian Transformation in Karnantaka.* Wageningen University Water Resources Series. New Delhi: Orient Longman.

Sinha, S., S. Gururani and B. Greenberg. 1997. "The 'New Traditionalist' Discourse of Indian Environmentalism," *Journal of Peasant Studies*, 24(3): 65–99.

Sillitoe, P. and A. Bicker 2003. "Introduction: Hunting for Theory, Gathering Ideology," in J. Pottier and P. Sillitoe (eds), *Development and Local Knowledge: New Approaches to Issues in Natural Resources Management*, pp. 1–18. London: Routledge.

White, S. 1996. "Depoliticising Development: The Uses and Abuses of Participation," *Development in Practice*, 6(1): 6–15.

Williams, R. 1973. *The Country and the City.* New York: Oxford University Press.

Winner, L. 1993. "Upon Opening the Black Box and Finding It Empty: Social Construction and the Philosophy of Technology," *Science, Technology and Human Values*, 18(3): 362–78.

About the Editors

Anjal Prakash is the Executive Director at South Asia Consortium for Interdisciplinary Water Resources Studies (SaciWATERs), Hyderabad, India. He is also the Project Director of "Water Security in Peri-Urban South Asia", a project funded by the International Development Research Centre (IDRC). He has worked extensively on the issues of groundwater management, gender, natural resource management, and water supply and sanitation. Having and advance degree from Tata Institute of Social Sciences (TISS), Mumbai, India and PhD in Social and Environmental Sciences from Wageningen University, The Netherlands, he has been working in the area of policy research, advocacy, capacity building, knowledge development, networking and implementation of large scale environmental development projects. Before joining SaciWATERs, he worked with the policy team of WaterAid India, New Delhi, where he handled research and implementation of projects related to Integrated Water Resources Management (IWRM). He is the author of *The Dark Zone: Groundwater Irrigation, Politics and Social Power in North Gujarat* (2005). His recent edited books are *Interlacing Water and Human Health: Case Studies from South Asia* (2012) and *Water Resources Policies in South Asia* (2013). He is presently co-editing two books on gender, water supply and sanitation concerns and peri urban water security issues.

Chanda Gurung Goodrich is Principal Scientist, Empower Women, International Crop Research Institute for the Semi-Arid Tropics (ICRISAT), Hyderabad, India. She holds a PhD from the Jawaharlal Nehru University, New Delhi, India. Her interest is in gender research in natural resource management, agriculture and livelihoods, especially for smallholder farmers. She has extensive experience in gender and participatory research and development, having worked as a researcher and consultant in the not-for-profit sector with various international and regional organizations in

South Asia, specializing in integrating social and gender equity into development programmes and projects. She also has published several papers in journals, and chapters in edited volumes on gender and agriculture/natural resource management.

Sreoshi Singh is Fellow, SaciWATERs. She specializes in urban development and planning and has a special interest in various socio-economic and environmental aspects of urban areas for policy planning. Besides, she has also worked on issues of poverty, livelihood, health and gender. She has been involved in the Crossing Boundaries Project of SaciWATERs since 2008 and presently leads the Hyderabad research team for the Peri-urban Water Security Project. She is currently pursuing her PhD from the Centre for Economic and Social Studies in Hyderabad, India and her topic of research is on the dynamics of water access for communities living in peri-urban Hyderabad.

Notes on Contributors

Sara Ahmed is Senior Program Specialist, International Development Research Centre (IDRC) Regional Office for South Asia and China, New Delhi, India. She has been a faculty member at the Institute of Rural Management, Anand, Gujarat, India, and an independent researcher and consultant working on the interface of water, social equity and climate change adaptation. Critically engaged with several civil society initiatives in South Asia, she was the chairperson of Gender and Water Alliance (GWA) (2007–09) at a time when women's voices, particularly from the South, were under-represented at global water forums. Her current work focuses on building research capacity to address the challenge of water and food security, particularly for women and vulnerable groups, in the context of climate variability and uncertainty in South Asia.

Sayeda Asifa Ashrafi holds a Master's degree in Water Resources Development from the Institute of Water and Flood Management at the Bangladesh University of Engineering and Technology, Dhaka. She has eight years of experience working with Bangladeshi and international organizations in increasingly challenging office and customer services management roles. She has in the past worked with the United Nations World Food Program in Dhaka and Action Aid Bangladesh. She also worked on program development through planning, policy analysis, interventions, project designing, community research and advocacy. She is presently based in Canada.

Partha Sarathi Banerjee is the Director of The Researcher, a civil society organization in Kolkata, West Bengal. Coming from the textile engineering background from Calcutta University, he has been working with the social sector for the last 15 years. He has worked extensively on both rural and urban water management in West Bengal. His fields of interest are irrigation, livestock, arsenic pollution, and groundwater policies in West Bengal. He has number of research publications in national and international journals to his credit.

Luisa Cortesi is a doctoral researcher at Yale School of Forestry and Environmental Studies, Yale University, New Haven, Connecticut, US, where she is studying the relation between environmental change, knowledge transmission and technologies of water management in flood-prone north Bihar. An environmental anthropologist, she is interested in social development processes, especially, equitable and sustainable water management. She has been in India since 2003, conducting extensive fieldwork in Tamil Nadu, working as State Coordinator for Megh Payan Abhiyan (MPA) in north Bihar (2007–09) and working for the United Nations. She is particularly concerned with bringing academic anthropological knowledge into the practice of social development and natural resource management.

Bhaskar Das is presently working as Associate Professor in VIT University, Vellore, Tamil Nadu. His areas of teaching and research are water resources management, hydrology, water and waste water quality monitoring and treatment. He has been in active research for last eight years and has authored and co-authored more than 30 international publications on groundwater arsenic and fluoride contamination in India and Bangladesh, groundwater market in West Bengal and socio-economic study.

Sanjukta Das is Reader in Economics at Sambalpur University, Sambalpur, Odisha, India. She has been a faculty member in the Department of Economics, Arunachal University (presently Rajiv Gandhi University), Arunachal Pradesh, and has participated in the preparation of Arunachal Pradesh Human Development Report as a member of the Technical Committee. Presently, she is engaged in a major research project funded by Indian Council of Social Science Research (ICSSR), New Delhi. Her areas of interest are poverty, inequality, human development and natural resource management. She has authored *Primary Education in Hills of Arunachal Pradesh: A Microstudy in India* (2011).

S. Janakarajan currently holds the position of Affiliate – Professor at the Madras Institute of Development Studies (MIDS), Chennai, Tamil Nadu, India and Professorial Research Associate at the School of Water and Development, SOAS, University of London. His areas of interest are development studies, rural development and agrarian institutions, climate change and adaptation, water management and irrigation institutions, urban and peri-urban water issues, conflicts and conflict resolution, water pollution, and groundwater management. He has published several books and articles in national and international journals. He is also the founder and convener of Cauvery Family, a platform to resolve the most vexed inter-state water dispute in the history of contemporary India between Karnataka and Tamil Nadu through farmer-to-farmer dialogue.

K. J. Joy is Senior Fellow, Society for Promoting Participative Ecosystem Management (SOPPECOM), Pune, Maharashtra, India. He has been an activist-researcher for more than 25 years and has a special interest in people's institutions for natural resource management both at the grassroots and policy levels. His other areas of interest include, drought and drought proofing, participatory irrigation management, river basin management and multi-stakeholder processes, watershed-based development, water conflicts and people's movements. He has been a visiting fellow to the Centre for Interdisciplinary Studies in Environment and Development (CISED), Bengaluru, Karnataka, India, and a Fulbright Fellow to the University of California, Berkeley. Presently he coordinates the national level initiative, "Forum for Policy Dialogue on Water Conflicts in India". He has published extensively and *Water Conflicts in India; A Million Revolts in the Making* (co-editor) is considered to be one of his important contributions.

Madhav Bahadur Karki is Senior Research Faculty at the Institute for Social and Environmental Transition, Nepal. Previously, he was Deputy Director General, International Centre for Integrated Mountain Development (ICIMOD), Kathmandu, Nepal. He has over 28 years of experience in the areas of natural resource management, network development in non-timber forest products (NTFPs) including, medicinal and aromatic plants (MAPs) in South Asia, and policy and partnership development. He has held various positions in several national and international organizations. He has authored and edited over 80 articles and books.

Nabendu Majumdar is a geophysicist at the Geological Survey of India, Kolkata, India. Previously, he has served as an Assistant Hydrogeologist at the Central Ground Water Board, Government of India for nine years. He has a PhD in "Geoelectric, Isotopic and Geochemical Studies for Hydrological and Chemical Characterization of Bakreswar Geothermal Area, Birbhum District, West Bengal" from Jadavpur University, Kolkata. He has published in journals of national and international repute, such as the *Journal of Applied Geophysics* and the *Journal of Applied Geochemistry*. His work has also been published in the proceedings of national and international symposia and conferences, such as the 2nd International Forum on Water and Food, held in Addis Ababa (Ethiopia), in 2008, co-hosted by ILRI and IWMI.

Nazeer Ahmed Memon is General Manager Transition, Sindh Irrigation and Drainage Authority (SIDA), Hyderabad, Sindh, Pakistan. A development professional with a long experience in implementing developmental and reform programs and in participatory irrigation management, he has contributed significantly towards decentralization and management transfer and organizing farmers' community and gender mainstreaming in water management. He has authored many articles, papers and edited books on water issues in Sindh.

Marcus Moench is President, Institute of Social and Environmental Transition (ISET), Boulder, Colorado, US. He has extensive experience of working with communities, governmental and non-governmental (both national and international) organizations on water, energy and forest management in South Asia, Middle East and western US. He combines a strong technical background in environmental science, hydrogeology and forestry with training and experience in the design and initiation of management institutions. He has led the India Water Sector Review, Groundwater Component and Yemen Decentralized Management Study for the World Bank. He also has published numerous journal articles on natural resources management.

Aditi Mukherji is the Theme Leader, Water and Air at ICIMOD, in Kathmandu, Nepal. Previously, she was a Senior Researcher at the International Water Management Institute (IWMI), based in Colombo and then in New Delhi. She has a PhD in Human Geography from the University of Cambridge and has worked in South Asia, Nile basin and Central Asia. She was the lead author of *Revitalizing Asia's Irrigation* – a joint effort of IWMI, Food and Agriculture Organization (FAO) and Asian Development Bank (ADB). She is an active blogger on water, agriculture and food issues since 2012. She was awarded the Inaugural Norman Borlaug award for Field Research and Application in 2012, endowed by the Rockefeller Foundation and given by the World Food Prize Foundation for her work on groundwater and electricity policies which led to change in groundwater law in her home state of West Bengal, India.

Vishal Narain is Associate Professor, School of Public Policy and Governance, Management Development Institute (MDI), Gurgaon, India. His teaching and research interests are in the analysis of public policy processes and institutions, water resource governance and peri-urban issues. He received the SR Sen Prize for the Best Book on Agricultural Economics and Rural Development (2002–03) conferred by the Indian Society for Agricultural Economics for his book, *Institutions, Technology and Water Control: Water Users Associations and Irrigation Management Reform in Two Large-Scale Systems in India* (2003); and has authored several articles in international refereed journals like *Water Policy, South Asian Water Studies*, etc. He has recently co-edited *Globalization of Water Governance in South Asia* (2014).

Suhas Paranjape is Senior Fellow, SOPPECOM. He has actively participated in different people's movements such as people's science movement, Adivasi agricultural labourers' movement, etc. He has been a core team member and consultant in many action research studies and pilot projects undertaken by SOPPECOM and Centre for Applied Systems Analysis in Development (CASAD), Pune, in the areas of participatory management of natural resources, especially in the field of irrigation. He has co-authored many books and co-edited *Water Conflicts in India: A Million Revolts in the Making* (2009).

Rezaur Rahman is Professor at the Institute of Water and Flood Management (IWFM), Bangladesh University of Engineering and Technology (BUET), Dhaka, Bangladesh. He holds a PhD in Environmental Engineering from University of Illinois, Urbana-Champaign, USA. At the national level, he has been involved in preparation of National Adaptation Program of Action (NAPA) for climate change and Climate Change Strategy and Action Plan (CCSAP) of Bangladesh.

Amita Shah is Director, Gujarat Institute of Development Research (GIDR), Ahmedabad, Gujarat, India. She is trained as an economist and has a wide-ranging experience of conducting research on various aspects of rural economy. Her major interests lie in natural resource development with a special focus on dry-land agriculture and forestry, environmental impact assessment (EIA), gender and environment, agriculture–industry interface, small-scale and rural industries, diffusion of technologies, and employment and livelihood issues. Recently, she has been involved in a number of studies pertaining to participatory watershed development, protected area management, economic valuation of biodiversity, chronic poverty in remote rural areas, and migration and status of women in agriculture. She has published about 100 research papers in journals and books.

Shaheen Ashraf Shah has a PhD in Women and Gender Studies from the University of Warwick, UK along with other credentials in the field of gender and development. She has worked as an independent gender and development researcher/consultant for range of national and international organizations such as the UNHCR Geneva, Asian Development Bank, USAID, among others. She has served as Gender and Governance Specialist for Gender Reform Action Plan and Gender Specialist for highly resistant Sindh water reforms program which has led to the mobilization of a large number of farmers, including women, in the Sindh province of Pakistan. Her research and professional interests include, democratization, women and gender rights, equity and inclusion issues, food security and climate change.

B. R. Sharma is Principal Researcher and Head, IWMI, New Delhi, India. He has over 25 years of experience in conducting and managing natural resource management research, especially for surface water and groundwater resources, on-farm water management, reclamation of saline and waterlogged soils, use of poor quality water, and development of water resources in hilly areas. His key contributions, include, among others, development of water savings techniques for irrigation of rice, improvement of on-farm water management in large irrigation commands, participatory irrigation management (PIM), hydrological sustainability of rice–wheat cropping system in Indo-Gangetic plains, development of regional salt, tillage and irrigation requirements of important crops, and watershed management.

Pranita Bhushan Udas is a doctoral researcher working on gendered participation in water management in Nepal at the Wageningen University, The Netherlands. Her interest lies in issues around gendered water equity and justice. Recently, she has carried out a study on women water professionals for SaciWATERs, as well as a study on the use of gender guidelines and manuals by water professionals in Nepal for GWA. At present she is a visiting faculty at Nepal Engineering College where she teaches a course on Gender and Social Inclusion and supervises Master level students on IWRM.

Ramesh Anand Vaidya is Senior Advisor, ICIMOD, Nepal, and leads research in the area of water storage for climate change adaptation in the Himalayan region. His research interests lie in the economics and policy of climate change and the consequences for water resources in the context of regional economic cooperation. He has taught at the University of Minnesota, Minnesota, US and has served as Nepal's Planning Commissioner for Water Resources and, later, Nepal's Ambassador to Japan, accredited concurrently to Australia. He has represented Nepal in several international water events and has published extensively on water resource issues, including recently on water-related disaster risk management, adaptation to climate change, and transboundary water resource management.

Index